Lecture Notes in Mathematics

History of Mathematics Subseries

Volume 2317

Series Editor

Patrick Popescu-Pampu, CNRS, UMR 8524 - Laboratoire Paul Painlevé, Université de Lille, Lille, France

Providing captivating insights into facets of the recent history of mathematics, the volumes in this subseries of LNM explore interesting developments of the past 200 years or so of research in this science. Their aim is to emphasize the evolution of its intellectual discourse, the emergence of new concepts and problems, the ground-breaking innovations, the human interactions, and the surrounding events that all contributed to weave the backdrop of today's research and teaching in mathematics. These high-level and largely informal accounts will be of interest to researchers and graduate students in the mathematical sciences, in the history or the philosophy of mathematics, and to anyone who seeks to understand the historical growth of the discipline.

Peter Roquette • Franz Lemmermeyer

The Hasse - Noether Correspondence 1925 - 1935

English Translation with Extensive Commentary

 Springer

Peter Roquette
Mathematical Institute
Heidelberg University
Heidelberg, Germany

Franz Lemmermeyer (iD)
Jagstzell, Germany

Translated by
Robert Perlis
Department of Mathematics
Louisiana State University
Baton Rouge, LA, USA

ISSN 0075-8434   ISSN 1617-9692   (electronic)
Lecture Notes in Mathematics
ISSN 2193-1771   ISSN 2625-7157   (electronic)
History of Mathematics Subseries
ISBN 978-3-031-12879-0   ISBN 978-3-031-12880-6   (eBook)
https://doi.org/10.1007/978-3-031-12880-6

Mathematics Subject Classification: Primary: 01A60, Secondary: 11-03

Edited and commented English translation of the original German edition published by Universitätsverlag Göttingen, 2006
Translation from the German language edition: "Helmut Hasse und Emmy Noether. Die Korrespondez 1925 - 1935" by Peter Roquette and Franz Lemmermeyer, © 2006. Published by Universitätsverlag Göttingen. All Rights Reserved.

This Springer imprint is published by the registered company Springer Nature Switzerland AG
The registered company address is: Gewerbestrasse 11, 6330 Cham, Switzerland

Preface

This book contains the English translation of all letters exchanged between Emmy Noether and Helmut Hasse as they are preserved in the *Handschriftenabteilung* of the Göttingen University Library. There are 82 such letters, dated from 1925 until Noether's sudden and tragic death in 1935.

The German text of these letters was published in the year 2006, together with Franz Lemmermeyer [LR]. The present translation is the response to widespread encouragement. I have supplemented the letters with detailed comments. These comments are not only the translation of our comments in the former German edition, but they are thoroughly reworked and updated. In addition I have added a new chapter with a general report on the development of class field theory in the years 1920–1940. Although Hasse and Noether were among the leading protagonists of this development, in their letters they discussed mainly certain details of the new viewpoints, as they arose during their work. Therefore it seems to me that it would be useful to be able to compare those details with the general line of development in their time. Perhaps this chapter may also be used as an introduction to class field theory, for those who are not yet familiar with it.

I would like to thank all people who helped me with the preparation of this volume, in particular the translator Robert Perlis for his insightful comments and his quick work. It was not an easy task to translate Noether's text into English without losing the flavor of her sometimes unconventional style.

All the letters cited in this book are contained in the Hasse legacy at the *Handschriftenabteilung* of the University Library in Göttingen (Cod. Ms. H. Hasse), except when explicitly referred to another source. The cited letters from Hasse to Davenport are contained in the Davenport legacy at Trinity College, Cambridge, England. The cited letters from Hasse to R. Brauer are contained in the Brauer legacy which is now at the *Handschriftenabteilung* in Göttingen too, thanks to the donation of Fred Brauer, son of Richard Brauer. The photo of Noether with Mr. and Mrs. Hasse comes from the archive of the Austrian Academy of Sciences, legacy of Auguste Dick, 13/1. The other photos in this book are from my private photo collection.

Heidelberg, Germany   Peter Roquette
May 2022

Contents

Chapter 1

Introduction

The name of one of the two correspondents, **Emmy Noether** (1882–1935), is known throughout the worldwide mathematical community.

She has been said to be

- *the creator of a new direction in algebra,*
- *the greatest woman mathematician who ever lived.*[1]

These are only some of the attributes which have been bestowed upon her in so many articles and speeches. A number of scientific institutions and projects carry her name as an icon. There is a serious Noether literature trying to understand and evaluate her impact on the development of mathematics that continues up to the present time. In the course of time her "new direction in algebra" has become standard not only in algebra but also

---

[1]Cited from Aleksandrov's address in memory of Emmy Noether, in Moscow on September 5, 1935, reprinted in English translation in the Collected Papers of Emmy Noether [Noe83b].

© Springer Nature Switzerland AG 2022
P. Roquette, F. Lemmermeyer, *The Hasse - Noether Correspondence 1925 - 1935*, History of Mathematics Subseries 2317,
https://doi.org/10.1007/978-3-031-12880-6_1

in general mathematical thinking, namely to work with mathematical structures, these structures being based on abstract axioms.

Emmy Noether 1930 (Foto by H. Hasse)

In a way our book can be viewed as part of this Noether literature, presenting a new and unique collection of Noetheriana. These Hasse-Noether letters are packed with mathematics, and they shed new light on how Emmy Noether arrived at her ideas and how she conveyed them to her peers. Through her open, unconventional and impulsive style, she allows us to have a glimpse not only into the working of her brain but also into her heart.

The other correspondent, **Helmut Hasse** (1898–1979), is known to the general mathematical public mainly through his Local-Global Principle. He has been said to be

- *one of the most important mathematicians of the twentieth century,*

- *a man whose accomplishments spanned research, mathematical exposition, teaching and editorial work.*[2]

The main field of Hasse's research was Number Theory. Apart from the already mentioned Local-Global Principle (for quadratic forms and for algebras) he contributed substantial results to class field theory (explicit reciprocity laws, norm theorem, local class field theory, complex multiplication), the theory of algebraic function fields (in particular with finite base fields where he succeeded in proving the Riemann hypothesis for elliptic curves). He wrote several successful and seminal books.

It may seem that these two mathematicians, Hasse and Noether, had somewhat different motivations and aims in their mathematical work. Whereas Hasse is remembered for his great concrete results in Number Theory, Emmy Noether's main claim to fame is not so much the theorems she proved but her methods. She herself has described her methods in one of her letters to Hasse (English translation):

> *My methods are working and conceptual methods and therefore have penetrated everywhere.*[3]

But on the other hand, their letters show that there existed a mutual understanding of the basic intellectual foundations of mathematical work, if not to say of its "philosophy". Both profited greatly from their contact. We see that not only did Hasse absorb Emmy's ideas on what was called "Modern Algebra" at that time, but conversely she became interested in the foundation of

---

[2]Cited from the Hasse-article by H. Edwards in the Dictionary of Scientific Biography, New York 1970–1990.

[3]This text can be found in the last paragraph of the letter of November 12, 1931. In the original German it reads: *"Meine Methoden sind Arbeits- und Auffassungsmethoden, und daher anonym überall eingedrungen."* We are not sure whether our translation faithfully reflects Noether's intentions but we have found no better alternative.

class field theory, to which she then contributed by proposing the use of non-commutative arithmetic as a powerful tool. In the course of time there arose a close cooperation of the two, and also a friendship as is evidenced from the tone of their letters.

Accordingly this book is to be regarded as a contribution not only to the literature on Noether but also to the literature on Hasse.

Unfortunately, the Hasse-Noether correspondence is preserved on one side only, i.e., the letters from Noether to Hasse. The letters in the other direction, i.e., from Hasse to Noether, are probably lost – except for three letters of which Hasse had made a copy for himself. Thus the Hasse-Noether correspondence, as available today, consists of 79 letters from Noether to Hasse and only 3 letters from Hasse to Noether. For some time we have tried to locate the missing letters from Hasse to Noether. After her sudden death 1935 in Bryn Mawr, at least part of Noether's legacy had been put into a container and sent to her brother Fritz who was living in Tomsk in Siberia at that time. Meanwhile we have found a letter from Fritz Noether to Hasse, dated October 2, 1935, from which we conclude that in all probability there are no further letters from Hasse that have been preserved.

Because of the lost Hasse letters, many parts of Noether's text which refer to his letters seem incomprehensible on first reading. However we have been able, by using other sources, to clear up most of the doubtful passages. Accordingly we have supplemented Noether's letters with detailed comments. In these comments we not only provide explanations in the technical sense but we also try to describe, in light of our present knowledge, the mathematical environment of Noether and that of Hasse, the mathematical tendencies of the time, and what was going on parallel to Hasse-Noether, as far as it is relevant to the text of the letters. Our comments appear immediately after the respective letter.

As already said, all of the Hasse-Noether letters (and postcards, respectively) are packed with mathematics. The correspondence partners did not only inform each other about their final results, but they also freely exchanged mathematical ideas and speculations, even when these could not yet be formulated in a precise manner, or when a convincing reason was still lacking. Noether called them "fantasies"; we would perhaps say "visions". Some of these visions have become reality and today belong to the basics of mathematics. Others have turned out not to be sustainable and had to be revised or abandoned. The Hasse-Noether correspondence is a rich source for those who are interested in the rise and the development of mathematical notions and ideas.

Not all the results and projects of Hasse, or those of Noether, are touched in their correspondence. Their letters are restricted to those topics which were close to their common interest. These include:

- Axiomatic algebra,

- Class field theory,

- Algebras and their arithmetics,

- Function fields.

In the 1920s and 1930s, those areas witnessed a particularly strong development whose effects can still be felt today. Emmy Noether and Helmut Hasse belonged to the outstanding protagonists of that development. From this viewpoint, the Hasse-Noether correspondence appears as a first rate historical document.

The exchange of letters between Hasse and Noether started in January 1925. At that time Hasse was 26 and held a position of *Privatdozent* at the University of Kiel. He had studied with Kurt Hensel in Marburg, and now strongly promoted the use of Hensel's p-adic numbers in Number Theory. While in Kiel, Hasse started a lively contact with Emil Artin in Hamburg, on explicit reciprocity laws and other topics of algebraic number theory; this continued until 1934. (The correspondence file of Artin-Hasse is published in translation in [FLR14].) In 1924 Hasse delivered a lecture course on class field theory in Kiel, of which there were lecture notes written by Reinhold Baer. Hasse's aim was to simplify and streamline the foundations of class field theory. In September 1925 at the meeting of the German Mathematical Society in Danzig (Gdansk), he delivered his famous survey on class field theory which finally led to his 3-volume class field theory report. In April 1925 Hasse was appointed full professor at the University of Halle. We believe that at that time he became the youngest professor at a German university.

Emmy Noether came to Göttingen in 1915, and since 1925 she held a position as *Privatdozent* with the official title of *"Außerordentlicher Professor"* (associate professor); this title did not carry a salary, and she was never promoted in Göttingen. In 1925 she could already look back on a successful mathematical career, in the sense that she had become a respected and highly valued member of the mathematical community. Originally she had worked under the influence of Hilbert and Klein, and her papers on differential invariants have become classic. In 1921 her seminal paper on ideal theory appeared in which she introduced and studied what today are called "Noetherian rings" on an axiomatic basis; more precisely: she showed that in a commutative ring the various decomposition laws of ideals into primary ideals (which were well-known for polynomial rings) can be obtained solely under the assumption that every ideal admits a finite basis. With this paper she started her own "completely original mathematical path" (Aleksandrov) into abstract algebra. At the 1924 meeting of the German Mathematical Society (DMV) in Innsbruck she presented the axiomatic basis for the factorization of ideals into prime ideals (known for rings of integers in an algebraic number field). Today those rings are called "Dedekind rings". Hasse was in attendance at that lecture and it left a great impression on him. At the meeting of the German Mathematical Society In September 1925, she delivered her influential talk in which she outlined her ideas on how to do representation theory, namely in the framework of the theory of abstract algebras. (This was at the same meeting and even in the same section in which Hasse delivered his survey on class field theory.)

In the first period of their correspondence, from 1925–1927, the topics of their letters do not seem to be particularly remarkable. They discussed Hasse's attempt to obtain an axiomatic description of unique factorization domains, there was some discussion of a possible axiomatic foundation of class field theory, and there were Noether's comments on how to present Galois theory in a textbook or in a lecture course.

In this first period the tone of Noether's letters sounds somewhat like that of a teacher to her student: criticizing and praising, giving good and not so good marks, encouraging and teaching. Noether, 16 years older than Hasse, was always ready to support young talent. We can observe that during this period, Hasse became increasingly interested in abstract axiomatic algebra as promoted by Emmy Noether.

This "teacher-student relationship" changed after 1927 when Noether asked Hasse about the existence of certain cyclic number fields. Within a few days Hasse was able to provide her with the construction of those fields. These were to be minimal splitting fields of high degree for the quaternion algebra. In his construction Hasse used (among other tools) the Local-Global Principle for quadratic forms which he had discovered; this was contained in his thesis. Emmy Noether was quite satisfied with Hasse's result. In a joint note with Richard Brauer on algebras and their splitting fields, she referred to Hasse's construction; Hasse's own note appeared immediately thereafter. These two notes together contain the first instance where the Local-Global Principle for some algebra was established, namely for the quaternion algebra.

Helmut Hasse

Subsequently Hasse became interested in the theory of algebras and their arithmetic, and Emmy Noether became more deeply interested in class field theory. In the following years a close collaboration arose among Noether, Hasse and R. Brauer. As a result Hasse developed the arithmetic of algebras by introducing and studying algebras over local fields. Shortly after, through his global theory of the norm residue symbol, he discovered local class field theory. Thus his study of algebras over local fields turned out to be a powerful tool for the foundation of local class field theory.

At the end of 1930, Hasse ventured to formulate his conjectures about simple algebras over number fields, including their Local-Global Principle and the conjecture that they are all cyclic. At first Noether did not believe in those conjectures but soon became convinced, and both started their quest for the proofs. This finally culminated in the famous Brauer-Hasse-Noether theorem, obtained at the end of 1931. This was an exciting year, and the excitement is mirrored in the Hasse-Noether correspondence.

Richard Brauer

Parallel to this we see the development of Noether's famous paper on integral bases for unramified extensions, and we learn of her motivation, her aim and her ideas around it.

In the second period 1927–1931, Noether addresses Hasse more and more as a colleague and partner on equal terms. In the third period, 1932–1935, the letters document the development of a genuine friendship.

This starts with Hasse's letter to Noether on the occasion of her 50th birthday on March 23, 1932. Her birthday was not officially observed at the University of Göttingen, but Hasse made sure that the algebraists in and outside of Göttingen were informed. He dedicated his paper on the structure of the Brauer groups of number fields to her, including a proof of Artin's reciprocity law of class field theory by means of algebras. In Noether's reply, and in all her subsequent letters we see her heartfelt affection towards her younger colleague. Mathematically, the later letters contain the attempts to generalize class field theory from abelian to arbitrary Galois extensions. Although these attempts did not reach their goal, it is of interest to note that the ideas of Noether, Artin and Hasse, were steps in the direction of what today is called algebraic cohomology theory.

Noether, Hasse and Frau Clark Hasse
This seems to be the only existing Foto showing Hasse and Noether
together

The friendship between Hasse and Emmy Noether stood its test in the hard times of summer 1933 after the Nazis had taken over the government in Germany. As documented in the letters, Hasse tried everything in his power to keep Noether in Göttingen. It is well-known that all this was in vain and she had to emigrate. But the friendly contact between Hasse and Noether continued, as we see in her letters from Bryn Mawr. As always, these letters were packed with mathematics. She never complained about her situation but in her letters we observe a sad undertone reflecting Noether's wish to return to her beloved Göttingen, which she had considered to be her home, the place where she had started her outstanding mathematical career and where she had risen to the height of her power.

The transcription and translation of the Noether letters turned out to be quite difficult, not only because Emmy used the old German handwriting of the 19th century, but also because of her really original style and her impulsive way of expressing herself. Most of the time she uses mathematical symbols without explanation, under the assumption (which was probably justified) that her correspondence partner will be able to figure out what is going on. Formulas and diagrams are condensed to tiny spaces on postcards with much additional text, which often is hardly decipherable. Except for some very few isolated spots we were finally able to read everything. Sometimes we wondered how Hasse managed to deal with her writing.

REMARK: In 1970, on the occasion of the 35th year after Emmy Noether's death, the short and empathetic biography of Emmy Noether by Auguste Dick was published [Dic70]. Looking through the pages of that biography one finds a number of citations from Noether's letters to Hasse. We conclude that Auguste Dick had access to those letters. But in 1970 Hasse was still alive and the Hasse legacy was not yet in the Göttingen library. So it seems that Dick had established personal contact to Hasse and that Hasse had shown her the letters. In fact, in the meantime we have found a letter from Auguste Dick to Hasse, dated September 7, 1967 where we read (English translation):

> *Respected Herr Professor, your mailing has surprised me and at the same time made me very happy. I do not know how to thank you. Already while preliminarily browsing through the 85 documents I have found much material which is quite important to me ... May I ask you to let me have those valuable documents until the beginning of November? ...*

We see that Hasse had sent her the whole collection of his correspondence with Emmy Noether. (And he sent the originals, for at that time making Xerox copies was not as common as it is today.) Thus all the letters published in this volume have already been in the hands of Auguste Dick. But apparently the possibility of publishing all the letters was not discussed when Hasse and Dick met in November 1967. She only asked for permission to cite those pieces of the text which are in her book. And Hasse granted this. Thus the Hasse-Noether letters had to wait many more years for their complete publication, and even more for their translation into English.

Chapter 2

Class Field Theory

Contents

The letters of Emmy Noether show that she was strongly involved in the evolution of modern class field theory, obviously inspired by Hasse. When we say "modern" class field theory then we have in mind the era starting with the seminal paper by the Japanese mathematian Teiji Takagi, which appeared in 1920 (in German) in a journal of Tokyo University [Tak20]. The main achievement of Takagi was to extend class field theory to include *all* abelian extensions of number fields. In earlier years the rules governing class fields had been observed only in certain special situations,

© Springer Nature Switzerland AG 2022
P. Roquette, F. Lemmermeyer, *The Hasse - Noether Correspondence 1925 - 1935*, History of Mathematics Subseries 2317,
https://doi.org/10.1007/978-3-031-12880-6_2

e.g., for field extensions connected with so-called complex multiplication (H. Weber), for quadratic extensions (Hilbert) and for cyclic extensions of prime degree (Furtwängler).

The first of Noether's letters to Hasse in which she mentions class field theory is that of November 17, 1926. She writes:

> *"I am glad that you continue to bring order into Tagaki. I notice more and more how your report simplifies the exposition; one only has to compare with Hilbert's one; how much effort is required there which is unnecessary today!"*

One year earlier, in September 1925, Noether had attended Hasse's one-hour-talk at the annual meeting of the DMV[1] in Danzig. There Hasse reported on Takagi's' new class field theory. Hasse had close contact to Emil Artin in Hamburg and both had realized the importance of Takagi's achievement.

Hasse's class field report was published in the year 1926 [Has26a].[2] We see from Noether's letter that she had read Hasse's report. It appears that she was impressed and became interested in class field theory. When she wrote that Hasse "continued" with Takagi, this indicates that she had heard that he was preparing an addition to his report. Indeed, at that time Hasse was working on such an addition. Whereas the original report presented only the main ideas, this addition, responding to demand, would contain full proofs. It appeared one year later [Has27a].

When Noether compares Hasse's report with Hilbert's famous *"Zahlbericht"* then she doesn't mean that Hasse's report should replace Hilbert's. Hasse had designed his report as a *sequel* to Hilbert's. In a letter from Hilbert to Hasse on November 5, 1926 Hilbert expresses his wish that Hasse write the report in such a way that everyone who is familiar with his (Hilbert's)

---

[1]DMV=*Deutsche Mathematiker Vereinigung*=German Mathematical Society.

[2]In German this became known as the *"Klassenkörperbericht."*

"Zahlbericht" could understand it "comparatively easily". Hasse makes this precise in the preface to his report by stating that he assumes chapters I–VII of Hilbert's report as known. Apparently Noether's criticism was aimed against the style of Hilbert's *Zahlbericht* [Hil97] which in her opinion was not sufficiently abstract.

The impact of Hasse's report was remarkable. Let us cite from a postcard of Bessel-Hagen to Hilbert, dated August 17, 1926, as an example of the reception of Hasse's class field report:

> *"A few days ago there appeared the latest issue of the DMV annual report which contains a report by Hasse on class field theory, and which is written in excellent clarity. The design of the whole theory is wonderfully uncovered by presenting the main ideas only while the proofs are reduced to their skeletons. Reading this article is a real pleasure; now all obstacles are eliminated which may have hampered access to the theory..."*

Formerly, class field theory had been the concern of only a select few specialists, but through Hasse's report it became known to a whole generation of mathematicians. Their names include Claude Chevalley, Max Deuring, Jacques Herbrand, Wolfgang Krull, Gottfried Köhte, Shokichi Iyanaga, Arnold Scholz, K. Shoda, Olga Taussky, Ernst Witt, Max Zorn and more, not to forget Emmy Noether. This had the effect that during the next two decades the proofs of class field theory became streamlined, simplified and shortened. The structure of abelian extensions of number fields became better understood.

Noether and Hasse took active part in this development; their letters provide ample witness for this. However, many of their letters deal with certain details of proofs; it is not easy to understand their role within the whole class field project of the time. Therefore, as a service to our readers, we shall present in this chapter a short sketch of class field theory in its development during the 1920s and 1930s, with special reference to the contributions of Hasse and Noether.

2.1   Takagi's Results

Let k be an algebraic number field of finite degree. The group D of its ideals contains the principal ideals H as a subgroup; the corresponding factor group D/H is called the group of *ideal classes*; this is a finite group. We assume the basic properties of algebraic number fields and their prime ideals to be known.

Teiji Takagi

Takagi's class field theory starts with a generalization of the above notion of ideal class: Let $\mathfrak{m}$ be an integer ideal of k. The group $D^{\mathfrak{m}}$ of ideals relatively prime to $\mathfrak{m}$ contains the subgroup $H^{\mathfrak{m}}$ of the principal ideals (a) which are generated by an element[3]

$$(2.1) \qquad a \equiv 1 \mod \mathfrak{m} \quad \text{and} \quad a \gg 0 \,.$$

$H^{\mathfrak{m}}$ is called the "ray modulo $\mathfrak{m}$". The corresponding factor group

$$C^{\mathfrak{m}} := D^{\mathfrak{m}}/H^{\mathfrak{m}}$$

is called the full group of *ray classes* modulo $\mathfrak{m}$. This is a finite group. Any factor group of $C^{\mathfrak{m}}$ is called a *ray class group modulo* $\mathfrak{m}$. If we wish to indicate which field k is considered then we add the index k and write $D_k^{\mathfrak{m}}, C_k^{\mathfrak{m}}$ etc.

Now let $K|k$ be a field extension of finite degree. Suppose every prime in $D^{\mathfrak{m}}$ is unramified in K. Consider the group of norms $N_{K|k}D_K^{\mathfrak{m}} \subset D_k^{\mathfrak{m}}$ and its image in the factor group $C_k^{\mathfrak{m}}$. Let us denote this image briefly as $N_K^{\mathfrak{m}}$. Thus the factor group $C_k^{\mathfrak{m}}/N_K^{\mathfrak{m}}$ is a ray class group modulo $\mathfrak{m}$. Takagi shows for its order:

$$(2.2) \qquad (C_k^{\mathfrak{m}} : N_K^{\mathfrak{m}}) \leq [K : k] \,.$$

This is called the "first fundamental inequality" of class field theory.[4] Takagi proves:

Theorem 1. *If equality holds in (2.2) then $K|k$ is abelian and its Galois group is isomorphic to the ray class group $C_k^{\mathfrak{m}}/N_K^{\mathfrak{m}}$. Moreover, the degree $f_{\mathfrak{p}}$ of any prime ideal $\mathfrak{p} \in D_k^{\mathfrak{m}}$ in K equals the order of $\mathfrak{p}$ in this ray class group.*

---

[3]$a \gg 0$ means that a is positive for every real archimedean valuation of k.

[4]The name "first" indicates that in Takagi's presentation it appears as the first important result. The "second" inequality says that for abelian extensions and only for those we have "$\geq$" in place of "$\leq$", hence equality. This statement is contained in Theorem 2. However at a later stage of the development, after Chevalley 1935, the arrangement of the proofs could be streamlined and the second inequality came first; accordingly the names of these inequalities were interchanged.

Recall that the degree $f_{\mathfrak{p}}$ is defined as the degree of the residue field extension:

$$(2.3) \qquad f_{\mathfrak{p}} = [K\mathfrak{P} : k\mathfrak{p}] = |G_{\mathfrak{p}}|$$

where $\mathfrak{P}$ denotes a prime divisor of K dividing $\mathfrak{p}$ and $|G_{\mathfrak{p}}|$ denotes the order of the decomposition group of $\mathfrak{P}$, i.e., the Galois group of $K\mathfrak{P}|k\mathfrak{p}$. Since $K|k$ is abelian, the decomposition group does not depend on the choice of the prime $\mathfrak{P}$ dividing $\mathfrak{p}$; therefore we have simply written $G_{\mathfrak{p}}$ and $f_{\mathfrak{p}}$. The number $f_{\mathfrak{p}}$ determines the decomposition type of $\mathfrak{p}$ in K. For, since $\mathfrak{p}$ is unramified, $\mathfrak{p}$ splits in K into a product of distinct primes:

$$\mathfrak{p} = \mathfrak{P}_1 \cdots \mathfrak{P}_r \quad \text{where} \quad r = \frac{[K : k]}{f_{\mathfrak{p}}}.$$

Hence, in the situation of Theorem 1 we see that the decomposition type in K of a prime $\mathfrak{p} \in D_k^{\mathfrak{m}}$ is determined by the class of $\mathfrak{p}$ in the ray class group $C_k^{\mathfrak{m}}/N_K^{\mathfrak{m}}$. Because of this property $K|k$ has been called "class field" with respect to $\mathfrak{m}$. This is the original definition of "class field" by Weber and Hilbert.

However, Takagi defined "class field" by the property

$$(2.4) \qquad (C_k^{\mathfrak{m}} : N_K^{\mathfrak{m}}) = [K : k].$$

In view of Theorem 1 we see that his definition is indeed equivalent to that of Weber and Hilbert. In addition to Theorem 1 Takagi's paper contains the following

Theorem 2. *Every finite abelian extension $K|k$ is class field with respect to a suitable integral ideal $\mathfrak{m}$ in k. Conversely, given an integral ideal $\mathfrak{m}$ in k and a ray class group modulo $\mathfrak{m}$ there exists a unique abelian extension $K|k$ which is class field modulo $\mathfrak{m}$ and the given ray class group is $C_k^{\mathfrak{m}}/N_K^{\mathfrak{m}}$.*

Such ideal $\mathfrak{m}$ is called a "module of definition" for the abelian extension $K|k$.

When Hecke reviewed Takagi's paper in the *Jahrbuch*[5] he called it "'final" (*abschließend*) in the sense that it completely solved the problem of describing the decomposition type of primes in finite abelian extensions of number fields by their behavior in the base field – at least for the unramified primes. But he observed that there still remainded the question of how to determine a suitable module of definition $\mathfrak{m}$ for a given abelian extension $K|k$. It turns out that such $\mathfrak{m}$ is not unique. The situation is as follows:

If $\mathfrak{m}$ divides $\mathfrak{m}'$ then there is a natural surjective map $C^{\mathfrak{m}'} \to C^{\mathfrak{m}}$, accordingly every ray class group modulo $\mathfrak{m}$ may be regarded as a ray class group modulo $\mathfrak{m}'$. More precisely, under this map the ray class group $C_k^{\mathfrak{m}'}/N_K^{\mathfrak{m}'}$ is mapped isomorphically onto $C_k^{\mathfrak{m}}/N_K^{\mathfrak{m}}$. It turns out that for a given abelian extension $K|k$ there is a unique smallest module of definition; this is called the "conductor" of $K|k$. Every multiple of the conductor is a module of definition for $K|k$. The conductor contains precisely the primes which are ramified in K. The exact description of the conductor for every given abelian extension $K|k$ was given in later years by Artin, Hasse and Noether.

If several abelian extensions of k are considered at the same time then it is convenient not to work with the conductor but to leave the choice of the module of definition open. Hasse in his class field report [Has26b] uses the following idea to construct a *unique* ray class group for a given abelian extension $K|k$: If a module of definition $\mathfrak{m}$ divides $\mathfrak{m}'$ then he identifies the two isomorphic ray class groups $C_k^{\mathfrak{m}}/N_K^{\mathfrak{m}}$ and $C_k^{\mathfrak{m}'}/N_K^{\mathfrak{m}'}$. Today we would interpret Hasse's identification procedure by taking the projective limit

$$(2.5) \qquad\qquad C^* = \varprojlim_{\mathfrak{m}} C^{\mathfrak{m}}$$

---

[5] The "*Jahrbuch für die Fortschritte der Mathematik*" was established in 1868 as the first comprehensive reviewing journal for the entire world literature in mathematics and related areas. It was active until 1945. Today the reviews of the *Jahrbuch* can be read and downloaded in the reviewing journal "zbMathOpen".

with respect to the maps $C^{m'} \to C^m$. This is a compact group, and Hasse's abstract ray class group for $K|k$ now appears as a finite factor group of C^*. Seen in this way Takagi's paper yields the following result:

Theorem 3. *The finite abelian extensions $K|k$ of the number field k correspond bijectively to the open subgroups N of C_k^* such that the Galois group $G(K|k)$ is isomorphic to the corresponding factor group C^*/N.*

Observe that at Hasse's time the notion of projective limit was not yet in use in number theory. The above formulation of Theorem 3 is today's interpretation of Takagi's results, which originally dealt only with *finite* abelian extensions. Whenever we talk about projective limits and their properties, we have to take into consideration that originally in the 1920s this was regarded as a statement on finite groups in the sense as explained above.

We have already said that Emmy Noether had read Hasse's class field report. In her letter of November 17, 1926 she encouraged him to continue in his endeavor to clarify the proofs of Takagi's results. The next letter is dated some weeks later. On December 11, 1926 she wrote:

> *It would be very nice if one could look closer into class field theory by way of axioms! Where have the things by F.K. Schmidt appeared, he has not yet sent them here!*

This question for axiomatization seems to us typical for Noether's way of thinking. In her seminal paper on what today is called "Noetherian rings" [Noe21] she had given proofs of the known facts about ideal theory of polynomial rings on an axiomatic basis. Similarly, her paper on "Dedekind rings" [Noe24] provided an axiomatization of the known ideal theory in number fields. Now she is asking for an axiomatization of class field theory. But when Noether mentions the name of F.K. Schmidt in this connection then this seems to be a misunderstanding.

F.K. Schmidt[6] (1901–1977) obtained his doctorate in Freiburg in the year 1925, officially with Loewy as his thesis advisor, but in reality advised by Wolfgang Krull who was *Privatdozent* in Freiburg at the time. Krull had spent several years in Göttingen and belonged to the circle around Emmy Noether. In the year 1925, shortly after receiving his doctorate, F.K. Schmidt went to the DMV-meeting in Danzig. There he attended Hasse's talk about Takagi's new class field theory. F.K. Schmidt was impressed and decided on the spot to try to transfer class field theory from number fields to function fields with finite base field. (Today we would say to "global fields of characteristic p".) It appears that Noether had heard of F.K. Schmidt's project, and she assumed that he was searching for a suitable system of axioms for Takagi's class field theory. But this was not the case. Hasse's legacy contains many letters from F.K. Schmidt, and one can see from them that his project was to transfer *analytic number theory* to the case of function fields with finite base field. He just tried to copy the proofs which he found in Hasse's class field report to the function field case. He was only partially successful in this endeavor; see, e.g., chapter 4 of [Roq18]. The full transfer of class field theory to global fields of characteristic p was later done by Hasse [Has34c], and in a different way by Witt [Wit34a]. The axiomatization of class field theory had to wait for many years. See, e.g., Artin-Tate [AT68] and Neukirch [Neu86].

The next letter of Noether, dated January 3, 1927, shows her continued interest; it contains some more ideas on the structure of class field theory, both for number fields and function fields.

---

[6]In Germany the name "Schmidt" is quite common, and there are a number of mathematicians with this name. Therefore it is usual to cite them along with their respective first name or first name initials.

2.2　Artin's Reciprocity Law

Let $K|k$ be a finite abelian extension of number fields. According to Takagi's Theorem 1 its Galois group $G(K|k)$ is isomorphic to the corresponding ray class group $C_k^{\mathfrak{m}}/N_K^{\mathfrak{m}}$, where $\mathfrak{m}$ is a module of definition in the sense explained above. But Takagi's proof of the existence of such an isomorphism was quite circuitous. He used analytic number theory and induction with respect to the field degree, and it would be hard if not impossible to extract from his proof an isomorphism in a canonical way. It was Artin who raised the question of a *canonical* construction of such isomorphism. In his paper [Art23] he conjectured and later in [Art27] he was able to prove:

Theorem 4 (Artin's Reciprocity Law). *The map sending every* $\mathfrak{p} \in D_k^{\mathfrak{m}}$ *to its Frobenius automorphism induces an isomorphism of the ray class group* $C_k^{\mathfrak{m}}/N_K^{\mathfrak{m}}$ *onto the Galois group* $G(K|k)$.

Recall the definition of the Frobenius automorphism: Let us denote by $|\mathfrak{p}|$ the number of elements in the residue field $k\mathfrak{p}$ (i.e., $|\mathfrak{p}|$ is the absolute norm of $\mathfrak{p}$). If $\mathfrak{P}$ is a prime divisor of $\mathfrak{p}$ in K then $|\mathfrak{P}| = |\mathfrak{p}|^{f_\mathfrak{p}}$. Recall that every $\mathfrak{p} \in D^{\mathfrak{m}}$ is unramified in K. The Frobenius automorphism $\left(\frac{K}{\mathfrak{p}}\right) \in G(K|k)$ is defined by the property

$$(2.6) \qquad a^{\left(\frac{K}{\mathfrak{p}}\right)} \equiv a^{|\mathfrak{p}|} \mod \mathfrak{P} \qquad \text{if } a \in K \text{ is } \mathfrak{P}\text{-integral.}$$

Since $K|k$ is abelian, $\left(\frac{K}{\mathfrak{p}}\right)$ is independent of the choice of $\mathfrak{P}$ as a prime divisor of $\mathfrak{p}$, and it is a generator of the cyclic decomposition group $G_\mathfrak{p}$.[7] For $\mathfrak{a} \in D^{\mathfrak{m}}$ the symbol $\left(\frac{K}{\mathfrak{a}}\right)$ is defined multiplicatively with respect to $\mathfrak{a}$.

It is clear that $\mathfrak{a} \to \left(\frac{K}{\mathfrak{a}}\right)$ is a homomorphism of the ideal group $D_k^{\mathfrak{m}}$ into the Galois group $G(K|k)$. This map is surjective according, e.g., to the Frobenius density theorem which was well-known at the time. This map vanishes on the norm group $N_{K|k}D_K^{\mathfrak{m}}$ since

---

[7]The decomposition group of $\mathfrak{P}$ does not depend on the choice of $\mathfrak{P}$ as a prime divisor of $\mathfrak{p}$ since $K|k$ is abelian. Hence we denote it by $G_\mathfrak{p}$.

$N_{K|k}\mathfrak{P} = \mathfrak{p}^{f_\mathfrak{p}}$ and $(\frac{K}{\mathfrak{p}})^{f_\mathfrak{p}} = 1$. Thus the essential statement of Artin's reciprocity law is that this map vanishes on principal divisors (a) with $a \equiv 1 \mod \mathfrak{m}$ and $a \gg 0$.

Artin's proof is not difficult; he compares the abelian extension $K|k$ with a suitable cyclotomic extension of k, where the contention of the theorem was known. This method is called "abelian crossing". Artin wrote in a letter to Hasse dated July 19, 1927 that he had learned this method from a paper by Chebotarev [Che26]. But otherwise, Artin's proof rested heavily on Takagi's class field results.

But why did Artin call his theorem "Reciprocity Law"? One would rather expect the name "Isomorphism Theorem" or something similar. The name "Reciprocity Law" has historical roots, connected with what is historically understood as "Reciprocity Law" for prime numbers. See, e.g., [Lem00].

Artin's Reciprocity Law can be written as a product formula for the so-called "norm residue symbol". Let us explain:

If $\mathfrak{p} \in D^\mathfrak{m}$ then $\mathfrak{p}$ is unramified in K and its norm residue symbol is defined for $0 \neq a \in k$:

$$(2.7) \qquad \left(\frac{a, K}{\mathfrak{p}}\right) := \left(\frac{K}{\mathfrak{p}}\right)^{v_p(a)}.$$

Here, $v_\mathfrak{p}(a)$ is the exponent with which $\mathfrak{p}$ appears in the principal divisor (a). Thus $v_\mathfrak{p}$ is the (additively written) valuation of k which belongs to the prime ideal $\mathfrak{p}$. The map $a \to (\frac{a,K}{\mathfrak{p}})$ is a surjective homomorphism of the multiplicative group $k^\times$ to the decomposition group $G_\mathfrak{p}$. If π is a prime element for $\mathfrak{p}$ then $(\frac{\pi,K}{\mathfrak{p}}) = (\frac{K}{\mathfrak{p}})$ is the Frobenius automorphism. Thus the map $a \to (\frac{a,K}{\mathfrak{p}})$ is essentially the $\mathfrak{p}$-part of the Artin map which gives Artin's Reciprocity Law. Its kernel consists of those $a \in k^\times$ for which $v_\mathfrak{p}(a) \equiv 0 \mod f_\mathfrak{p}$. Since the residue field $k\mathfrak{p}$ is finite each of its elements is a norm from $K\mathfrak{P}$; hence every element $a \in k^\times$ is a norm residue modulo arbitrary sufficiently high powers of $\mathfrak{p}$; this means that for every sufficiently large r there exists $\alpha_r \in K^\times$ such that[8]

$$a \equiv N_{K|k}(\alpha_r) \mod \mathfrak{p}^r.$$

(It suffices that $r \geq f_{\mathfrak{p}}$.) In view of this property the symbol $\left(\frac{a,K}{\mathfrak{p}}\right)$ is called the "norm residue symbol" with respect to $\mathfrak{p}$.

The question arises whether for prime ideals $\mathfrak{q}|\mathfrak{m}$ there also exists such a norm residue symbol, i.e., a canonical surjective homomorphism $a \to \left(\frac{a,K}{\mathfrak{q}}\right)$ of $k^\times$ to the decomposition group $G_{\mathfrak{q}}$ whose kernel consists of those elements which are norm residues modulo arbitrary sufficiently high powers of $\mathfrak{q}$. This is indeed the case. But the definition cannot be given as in (2.7) since a prime divisor $\mathfrak{q}$ of $\mathfrak{m}$ may be ramified in K and hence the Frobenius automorphism is not defined as an element in $G(K|k)$. But Hasse had found a method to define $\left(\frac{a,K}{\mathfrak{q}}\right)$ also for $\mathfrak{q}|\mathfrak{m}$ [Has30d]. He proved:

Theorem 5. *For every prime $\mathfrak{p}$ of k, whether $\mathfrak{p}|\mathfrak{m}$ or $\mathfrak{p} \in D^{\mathfrak{m}}$, there exists a unique norm residue symbol $\left(\frac{a,K}{\mathfrak{p}}\right)$ in the above sense, which for unramified $\mathfrak{p}$ is given by (2.7), such that the product formula holds:*

$$(2.8) \qquad\qquad \prod_{\mathfrak{p}} \left(\frac{a,K}{\mathfrak{p}}\right) = 1$$

if $a \gg 0$. Note that for each $a \in k^\times$ this is a finite product, for in view of (2.7) there are only finitely many $\mathfrak{p} \in D^{\mathfrak{m}}$ with $\left(\frac{a,K}{\mathfrak{p}}\right) \neq 1$.

On first sight this looks quite different from Artin's Reciprocity Law but a closer look will show that both statements are essentially the same. To see this, write (2.8) in the form

$$(2.9) \qquad\qquad \prod_{\mathfrak{q}|\mathfrak{m}} \left(\frac{a,K}{\mathfrak{q}}\right) = \prod_{\mathfrak{p} \in D^{\mathfrak{m}}} \left(\frac{a,K}{\mathfrak{p}}\right)^{-1}$$

if $a \gg 0$. The right hand side is the inverse of the Artin map, and hence vanishes on those $a \in k^\times$ for which $a \equiv 1 \mod \mathfrak{m}$

---

[8]Here and in the following, congruences are meant in the multiplicative sense, i.e., the congruence above means $\frac{a}{N\alpha_r} \equiv 1 \mod \mathfrak{p}^r$.

and $a \gg 0$. Therefore the left hand side vanishes too for these a. Since the finitely many primes $\mathfrak{q}|\mathfrak{m}$ and the infinite primes are independent, for given $a \in k^\times$ and $\mathfrak{q}|\mathfrak{m}$ there exists an auxiliary $a' \in k^\times$ such that

$$a' \equiv a \mod \mathfrak{q}^r,$$
$$a' \equiv 1 \mod \mathfrak{q}'^r \text{ for all } \mathfrak{q}' \neq \mathfrak{q}, \quad \mathfrak{q}'|\mathfrak{m}$$

and $a' \gg 0$, where the exponent r is sufficiently large. (It suffices that r is greater than the exponents with which $\mathfrak{q}$ and the $\mathfrak{q}'$ appear in $\mathfrak{m}$.) Then $\left(\frac{a', K}{\mathfrak{q}'}\right) = 1$ for $\mathfrak{q}' \neq \mathfrak{q}$ and (2.9) reads

$$(2.10) \qquad \left(\frac{a, K}{\mathfrak{q}}\right) = \left(\frac{a', K}{\mathfrak{q}}\right) = \prod_{\mathfrak{p} \in D^\mathfrak{m}} \left(\frac{a', K}{\mathfrak{p}}\right)^{-1}.$$

In view of Artin's Reciprocity Law the right hand side does not depend on the auxiliary element a'. This shows that for every $\mathfrak{q}|\mathfrak{m}$ the norm residue symbol $\left(\frac{a, K}{\mathfrak{q}}\right)$ is uniquely determined by the product formula (2.8). Now Hasse used (2.10) as a *definition* of $\left(\frac{a, K}{\mathfrak{q}}\right)$ and then he verified, using Artin's Reciprocity Law, that indeed this is a norm residue symbol in the sense explained above.

Artin proved his Reciprocity Law during a lecture course in Hamburg in the summer of 1927. Immediately after that he informed Hasse about it. In a letter of July 19, 1927 he asked whether Hasse would now be able to develop the product formula. Hasse did so, but his paper [Has30c] appeared only three years later, in 1930. This delay can be explained. The product formula brought a surprise: Hasse discovered *local* class field theory, and this led to a completely new aspect which had to be taken care of.

Before explaining this, let us remark that for the infinite primes, i.e., the archimedean valuations of k, there also exist norm residue symbols. Although these are rather trivial, it is convenient to include them into the formula (2.8) since then the side conditions like $a \gg 0$, which refer to the infinite real primes, can be dropped. An infinite prime $\mathfrak{p}_\infty$ is either real or complex. Let $\mathfrak{P}_\infty$ be an extension of $\mathfrak{p}_\infty$ to K. If $\mathfrak{P}_\infty$ and $\mathfrak{p}_\infty$ are both real or both complex then the decomposition group $G_{\mathfrak{p}_\infty} = 1$ and the corresponding norm symbol $\left(\frac{a,K}{\mathfrak{p}_\infty}\right)$ is trivial. Otherwise $\mathfrak{p}_\infty$ is real and $\mathfrak{P}_\infty$ is complex; the decomposition group $G_{\mathfrak{p}_\infty}$ is of order 2, generated by the corresponding complex conjugate automorphism. Let us denote this by $\left(\frac{K}{\mathfrak{p}_\infty}\right)$. Then $\left(\frac{a,K}{\mathfrak{p}_\infty}\right) = \left(\frac{K}{\mathfrak{p}_\infty}\right)$ if $a < 0$, and $= 1$ if $a > 0$. With these definitions, the product formula (2.8) holds if $\mathfrak{p}$ ranges over *all* primes of k, finite or infinite, and for *all* $a \in k^\times$, without any side conditions.

Emil Artin and Helmut Hasse 1930

2.3   Hasse's Local Class Field Theory

Let k be a number field and $\mathfrak{p}$ a fixed prime ideal of k. Denote by $k_\mathfrak{p}$ the $\mathfrak{p}$-adic completion of k. Consider a finite abelian extension $K|k$ and let $\mathfrak{P}$ be an extension of $\mathfrak{p}$ to K. The $\mathfrak{P}$-adic completion of K is a finite abelian extension of $k_\mathfrak{p}$; as such it does not depend on the choice of $\mathfrak{P}$ as a prime divisor of $\mathfrak{p}$ in K, so let us denote it by $K_{\mathfrak{p}}$. The Galois group of $K_{\mathfrak{p}}|k_{\mathfrak{p}}$ is naturally isomorphic to the decomposition group $G_{\mathfrak{p}}$ and can be identified with it.

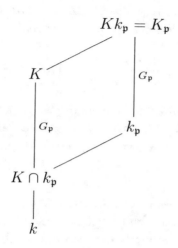

Hasse's norm residue symbol, which is a homomorphism $a \to \left(\frac{a,K}{\mathfrak{p}}\right)$ from $k^{\times}$ onto $G_{\mathfrak{p}}$, extends by $\mathfrak{p}$-adic continuity to a homomorphism from $k_{\mathfrak{p}}^{\times}$ onto $G_{\mathfrak{p}}$. The kernel of the latter is the closure of the kernel of the former which, in view of Theorem 5, consists of those elements in $k^{\times}$ which are norm residues from K modulo arbitrary high powers of $\mathfrak{p}$. Because of $\mathfrak{p}$-adic completeness such a norm residue is actually a norm from $K_{\mathfrak{p}}$. In this way, through the passage to the complete field, the "norm residue symbol" becomes the "norm symbol". In other words: The map $a \to \left(\frac{a,K_{\mathfrak{p}}}{\mathfrak{p}}\right)$ induces an isomorphism from the norm factor group $k_{\mathfrak{p}}^{\times}/NK_{\mathfrak{p}}^{\times}$ to the Galois group $G_{\mathfrak{p}}$ of $K_{\mathfrak{p}}|k_{\mathfrak{p}}$.

Hasse observed these facts in his paper [Has30c]. At the same time he conjectured that for every given finite abelian extension of the local field $k_{\mathfrak{p}}$ there exists a finite abelian extension K of the global field k whose completion $K_{\mathfrak{p}}$ coincides with the given extension of $k_{\mathfrak{p}}$. Moreover, for any given finite factor group of $k_{\mathfrak{p}}^{\times}$ there exists a finite abelian extension of $k_{\mathfrak{p}}$ such that its norm factor group coincides with the given factor group of $k_{\mathfrak{p}}^{\times}$. These conjectures were confirmed by F.K. Schmidt [Sch30] with whom Hasse held a close scientific correspondence. This gives the following

Theorem 6 (Local Class Field Theory). *Let $k_\mathfrak{p}$ be the localization of a number field at a prime ideal $\mathfrak{p}$. For every finite abelian extension $K_\mathfrak{p}$ of $k_\mathfrak{p}$ the $\mathfrak{p}$-adic norm symbol*

$$a \to \left(\frac{a, \, K_\mathfrak{p}}{\mathfrak{p}} \right) \qquad (for \ a \in k_\mathfrak{p}^\times)$$

induces an isomorphism of the norm factor group $k_\mathfrak{p}^\times / N K_\mathfrak{p}^\times$ to the Galois group $G(K_\mathfrak{p}|k_\mathfrak{p})$. This establishes a 1-1 correspondence of the finite abelian extensions of $k_\mathfrak{p}$ with the finite factor groups of $k_\mathfrak{p}^\times$.

Whereas in the "global" class field theory the finite abelian extensions correspond to the ray class groups, now in the local case the finite abelian extensions correspond to the norm class groups of the base field.

Artin's Reciprocity Law for a global abelian extension $K|k$ can be obtained from the local norm symbols by means of the product formula (2.8) for $a \in k^\times$.

We have already said that this discovery came to Hasse as a surprise, although the process of localization is quite straightforward and not at all complicated, once Artin's reciprocity law had been established as product formula (2.8). In our opinion, this discovery is to be rated as one of the most important events of number theory in the twentieth century. It shows that one of the main achievements of number theory, namely the description of abelian extensions due to Takagi and Artin, can be obtained from the local norm symbols via their product formula. This established the local fields definitely as indispensable objects of number theory, implementing the original idea of Hensel who had discovered the local fields and introduced them into number theory.

But this is not the end of the story. For these new viewpoints were still based on Takagi's paper whose proofs are long-winded, and it is not easy to detect a structure there. As said already, in his class field report Hasse had started to organize Takagi's work. He was looking for a comprehensible structure in class field theory. Now finally he found it: the path from the local norm symbols to the global Reciprocity Law. But this path had to be cleared, and so started the quest for an independent foundation of the local theory, and thereafter the combination of all local components to obtain the global product formula.

Let us cite from Hasse's paper [Has30c] (in English translation):

> *I expect that a construction of class field theory which, conversely, first develops the theorems of local class field theory* ab ovo *and then crosses the border to global class field theory – perhaps by means of the Prüfer-v.Neumann theory of ideal numbers – would give a considerable conceptual, if not factual, simplification of the proofs of global class field theory, which in their hitherto rough state do not make the study of this theory attractive, although it is so appealingly smooth in its results.*

And so the project of building class field theory through its local components started.

2.4 Noether's Algebras: Local Case

Up to now we have seen Emmy Noether as an interested observer of the new class field theory, but not yet as an active contributor. This changed as soon as Hasse had proposed his above mentioned project. Her main contribution to Hasse's project was to propose using *non-commutative algebras*.[9] This is explicitly mentioned in Hasse's paper [Has33b] which he had dedicated to Emmy Noether on the occasion of her 50th birthday.

---

[9]Noether herself always used the old-fashioned term "hypercomplex system" instead of "algebra".

The origin of Noether's occupation with algebras goes back to the year 1925. Recall that in 1925 Hasse delivered his report on Takagi's class field theory at the annual DMV-meeting in Danzig. At the same meeting, and even in the same section Emmy Noether delivered a talk about her new ideas on representation theory. (The section was chaired by Kurt Hensel.) Noether presented her idea to look at group representations not as matrices but rather as operator modules with respect to a group – or with respect to the corresponding group algebra. These ideas were later expanded in her lectures and discussions. At some point she cooperated with Hasse on a question about splitting fields of an algebra, but this had no direct relation yet to class field theory (see postcard of October 4, 1927). Noether's ideas found a comprehensive presentation in her important paper [Noe29]. That paper has been called a "pillar in linear algebra" (Curtis). Well, although Noether is correctly listed as the author, the paper was actually written by her student Van der Waerden who had taken notes on her lecture in the winter term 1927/28. This reflects the nature of Noether's extraordinary influence on the development of mathematics in her time: she did not publish much herself but she spread her ideas freely, and she encouraged other people to elaborate on them and to publish. Today, modules with operators belong to the standard tools of a mathematician; but this was not always the case and goes back to Emmy Noether.

During her studies of representations Noether had to develop the structure of algebras. Actually, algebras had been studied earlier in the USA, mainly by Wedderburn and Dickson. The German translation of Dickson's book [Dic23] started an intensive development of the structure theory of algebras, including their arithmetic properties. A comprehensive presentation of the theory of algebras in Noether's time can be found in the book by Deuring [Deu35a]. Originally Noether herself had been asked to write a book on algebras, in the newly established book series *"Ergebnisse der Mathematik"* of Springer Verlag. But she declined and proposed that her student Max Deuring should write such a book. And Deuring did so [Deu35a]. In Noether's letter to Hasse of November 8, 1931 Sect. 3.34 she says that Deuring's book is planned to be finished in May 1932. However, the book appeared only in 1935. The delay was due to the rapid development of the theory of algebras in those years, in connection to class field theory. Noether took an active part in the preparation of the book. Even after she was forced to emigrate to the USA she corresponded frequently with Deuring about it. In her letter of November 23, 1934 Noether informed Hasse that Deuring had finally completed the manuscript for his book. This was several months before Noether's sudden death. In our opinion, Deuring's book is the best source on the theory of algebras in Noether's time.

Max Deuring

Helmut Hasse was well informed about the development of Noether's theory of algebras, when he started to use it in his new foundation of local class field theory. One can say that without his close scientific contact to Emmy Noether he would not have been able to find his local construction of the norm symbol $\left(\frac{a,K}{\mathfrak{p}}\right)$.

In local class field theory which Hasse had discovered, the norm factor group plays a fundamental role (see the preceding section). Perhaps this was the reason why Hasse now, in his next paper [Has31d], was mainly interested in *cyclic* algebras which, by definition, are connected to norm factor groups. Consider a cyclic field extension $K|k$ of degree n, let σ be a generator of the Galois group G, and $0 \neq \alpha \in k$. The corresponding cyclic algebra $\mathcal{A}$ contains K and an element u with the *defining relation*

$$au = ua^\sigma \quad \text{for all } a \in K$$
$$u^n = \alpha.$$

This is a central[10] simple algebra over k of k-dimension n^2. It is called a *"crossed product"* of the cyclic field $K|k$ with its Galois group. Notation:

$$\mathcal{A} = (\alpha, K, \sigma).$$

Since the time of Dickson it is well-known that this is a division algebra if and only if α^n is the smallest power of α which is a norm from K. In this case, the map $\alpha \to \sigma$ generates an isomorphism of the norm factor group $k^\times/NK^\times$ with the Galois group G of $K|k$.

As an example, let W_n be the unramified extension field of the $\mathfrak{p}$-adic field $k_\mathfrak{p}$ of degree n. W_n is cyclic, and its Galois group is generated by the Frobenius automorphism $\left(\frac{W_n}{\mathfrak{p}}\right)$. Let π be a prime element and consider the cyclic $k_\mathfrak{p}$-algebra

$$(2.11) \qquad\qquad \mathcal{W}_n := \left(\pi, W_n, \left(\frac{W_n}{\mathfrak{p}}\right)\right).$$

---

[10]The attribute "central" means that the base field k is the center of the algebra. In Noether's time one said "normal" instead of "central" in this connection, at least in the German literature.

This is a division algebra since π^n is the first power of π which is a norm from W_n.

Hasse's main result in the local case is

Theorem 7. *Let $\mathcal{D}$ be a central division algebra over $k_{\mathfrak{p}}$ of index n.[11] Every extension field K of $k_{\mathfrak{p}}$ of degree n is a splitting field [12] of $\mathcal{D}$, hence can be isomorphically embedded into $\mathcal{D}$. Such an isomorphism is unique up to an inner automorphism of $\mathcal{D}$.*

Hasse's proof combines arguments of valuation theory of skew fields with those of the theory of algebras.

Now let $K|k_{\mathfrak{p}}$ be a cyclic field extension of degree n and σ a generating automorphism of its Galois group G. Applying the theorem to the division algebra W_n we may suppose $K \subset W_n$. Every automorphism of K can be extended to an automorphism of W_n. Every automorphism of a central simple algebra is an inner automorphism. Hence there exists $u \in W_n$ such that $u^{-1}xu = x^{\sigma}$ for all $x \in K$. It follows that $\alpha := u^n$ is contained in the center of W_n, hence $\alpha \in k_{\mathfrak{p}}$. This establishes W_n as a cyclic crossed product of K with its Galois group:

$$(2.12) \qquad\qquad W_n = (\alpha, K, \sigma).$$

Since W_n is a division algebra it follows that α^n is the first power of α which is a norm from K. Hence the norm factor group $k_{\mathfrak{p}}^{\times}/NK^{\times}$ is isomorphic to the Galois group G, an isomorphism is given by $\alpha \mapsto \sigma$. This map does not depend on the choice

---

[11]The $k_{\mathfrak{p}}$-dimension of $\mathcal{D}$ is a square, its square root is called the "index" of $\mathcal{D}$.

[12]Recall that a field K is said to split $\mathcal{D}$ if the tensor product $\mathcal{D} \otimes K$ is a full matrix algebra over K.

of the generating element σ of G.[13] We denote it like Hasse as $a \rightarrow (\frac{a,K}{\mathfrak{p}})$. More precisely:

(2.13)   if $a \equiv \alpha^r \mod NK^{\times}$ then $\left(\dfrac{a, K}{\mathfrak{p}}\right) = \sigma^r$.

Of course, the exponent r is relevant in its residue class modulo n only.

The local construction of the norm symbol given above is canonical and it turns out to be compatible with extending the field K, as expected. If $K = W_n$ then in view of (2.11) we obtain the Artin map defined by $\pi \mapsto (\frac{W_n}{\mathfrak{p}})$ which we have seen earlier in (2.7).

When Hasse informed Noether about his local construction of the norm symbols she replied: *"I have read your theorems with great enthusiasm, like an exciting novel"* (letter of April 12, 1931). Her excitement was triggered by the fact that she had expected this one year earlier (see her letter of June 25, 1930).

However the above construction works only for *cyclic* field extensions $K|k_{\mathfrak{p}}$. In order to establish the full local class field theory one has to consider arbitrary abelian extensions of $k_{\mathfrak{p}}$. In such a case one has to work not only with cyclic algebras but with algebras given by factor systems. This had been settled by Chevalley in the paper [Che33a] which appeared in Crelle's Journal. But he had already announced it in the year 1931 [Che31]. Moreover, Chevalley showed that every finite factor group of $k_{\mathfrak{p}}^{\times}$ is the norm factor group of a unique abelian field extension. It appears that Hasse had informed Noether about Chevalley's results, for in her letter of June 3, 1932 she asked Hasse to send her Chevalley's announcement.

---

[13]If σ is replaced by σ^{ν} with ν relatively prime to n then u is to be replaced by u^{ν} and hence α by α^{ν}.

Claude Chevalley

By the way, Hasse's Theorem 7 can also be applied to embed W_n into $\mathcal{D}$, showing that every division algebra $\mathcal{D}$ of index n can be written as a cyclic crossed product:

$$(2.14) \qquad\qquad \mathcal{D} = (\beta, W_n, (\frac{W_n}{\mathfrak{p}})).$$

If $v_{\mathfrak{p}}(\beta) = r$ then the quotient $\frac{r}{n}$ modulo 1 is an invariant of $\mathcal{D}$, called the "Hasse invariant". This leads to a description of the Brauer group[14] over $k_{\mathfrak{p}}$: it is isomorphic to the group of rational numbers modulo 1. Moreover, this implies that the exponent of a central simple algebra over the local field $k_{\mathfrak{p}}$ equals its index.

---

[14]The Brauer group of $k_{\mathfrak{p}}$ consists of the similarity classes of central simple algebras over $k_{\mathfrak{p}}$. Two algebras $\mathcal{A}, \mathcal{B}$ are called "similar" if they are matrix algebras over the same division algebra. Notation: $\mathcal{A} \sim \mathcal{B}$. The group operation in the Brauer group is given by the tensor product $\mathcal{A} \otimes \mathcal{B}$. (In Noether's time one wrote $\mathcal{A} \times \mathcal{B}$ for the tensor product.) The "'exponent"' of $\mathcal{A}$ is the smallest positive number e such $\mathcal{A}^e$ splits.

2.5   Local to Global

Now we work over an algebraic number number field k. Let $\mathcal{A}$ be a central simple algebra over k. For any prime ideal $\mathfrak{p}$ of k the algebra

$$\mathcal{A}_{\mathfrak{p}} := \mathcal{A} \otimes k_{\mathfrak{p}}$$

is a central simple algebra over $k_{\mathfrak{p}}$. There are only finitely many $\mathfrak{p}$ for which $\mathcal{A}_{\mathfrak{p}}$ does not split; this is seen through a suitable definition of the different of a central simple algebra. The first question is whether $\mathcal{A}$ is uniquely determined by its $\mathfrak{p}$-adic components $\mathcal{A}_{\mathfrak{p}}$, at least up to similarity. This is equivalent to the famous

Theorem 8 (Local-Global Principle). *If each local $\mathfrak{p}$-component $\mathcal{A}_{\mathfrak{p}}$ splits over $k_{\mathfrak{p}}$ then $\mathcal{A}$ splits over k.*

Here one has to include the infinite primes $\mathfrak{p}_{\infty}$ of k, as we had already seen in Artin's Reciprocity Law. If $k_{\mathfrak{p}_{\infty}}$ is real then the algebra $\mathcal{A}_{\mathfrak{p}_{\infty}}$ is similar either to the quaternion algebra $\mathbb{H}$, or it is split and $\mathcal{A}_{\mathfrak{p}_{\infty}} \sim \mathbb{R}$.

The Local-Global Principle is part of Hasse's conjectures which he had written to Noether and which she did not believe at first; see her letter of December 12, 1930. But soon she changed her mind and she and Hasse united to search for a proof. After a long exciting story, which can be followed in their correspondence, Noether finally discovered a proof which she called "trivial". You can find her proof in the postcard of November 11, 1931.

Noether's argument is indeed trivial, but it should be noted that it is just a reduction to the case when $\mathcal{A}$ is split by a cyclic field $K|k$ of prime degree. In the cyclic case the theorem can be formulated as follows:

Theorem 9 (Norm theorem). *Let $K|k$ be cyclic. If an element $\alpha \in k$ is a norm from $K_{\mathfrak{p}}$ for every prime $\mathfrak{p}$ then α is a norm from K.*

In the cyclic case of prime degree the norm theorem was well-known. It had been obtained by Furtwängler in the year 1912 and had been used by Takagi (and Artin). Of course Furtwängler did not consider $\mathfrak{p}$-adic fields but he worked in the number field k and considered "norm residues" modulo high powers of $\mathfrak{p}$ instead of norms from $k_{\mathfrak{p}}$. This was in the style of Hilbert who had first pointed out the importance of norm residues. But due to Hensel this is actually the same as working with $\mathfrak{p}$-adic norms.

Thus Noether's "simple" proof of the local-global principle rested fully on the old theorem of Furtwängler and hence did not completely fit into her project to give a completely *new* foundation of class field theory. But for the time being this was accepted by her and by Hasse.

The next conjecture of Hasse which he and Noether attacked was whether every central simple algebra over k is cyclic. This was suggested by the fact that this is true in the local case; see (2.14). (For infinite primes $\mathfrak{p}_\infty$ this is trivial since the quaternion algebra is cyclic.) In other words: If $\mathcal{A}$ is a central simple algebra, does there exist a cyclic splitting field for $\mathcal{A}$?

If K splits $\mathcal{A}$ then $K_{\mathfrak{p}}$ splits $\mathcal{A}_{\mathfrak{p}}$, for every prime $\mathfrak{p}$ of k. In the local case, this requires that the field degree $[K_{\mathfrak{p}} : k_{\mathfrak{p}}]$ is a multiple of the index $n_{\mathfrak{p}}$ of $\mathcal{A}_{\mathfrak{p}}$. This is a condition only for the finitely many $\mathfrak{p}$ for which $\mathcal{A}_{\mathfrak{p}}$ does not split.

Lemma 10. *Let S be a finite set of primes of k (infinite real primes allowed). For each $\mathfrak{p} \in S$ let $n_{\mathfrak{p}} \geq 1$ be a positive integer (for an infinite real prime $n_{\mathfrak{p}} \leq 2$). Then there exists a cyclic extension field $K|k$ such that $[K_{\mathfrak{p}} : k_{\mathfrak{p}}] = n_{\mathfrak{p}}$ for each $\mathfrak{p} \in S$. Moreover, K can be chosen such that its degree is a given multiple of all $n_{\mathfrak{p}}$.*

Accordingly, let S contain those primes for which $\mathcal{A}_{\mathfrak{p}}$ does not split, and let $n_{\mathfrak{p}}$ be the index of $\mathcal{A}_{\mathfrak{p}}$. Let $K|k$ be cyclic as in the lemma and consider $\mathcal{A} \otimes K$. Its $\mathfrak{p}$-component $\mathcal{A}_{\mathfrak{p}} \otimes K_{\mathfrak{p}}$ for $\mathfrak{p} \in S$ splits because of the degree condition in the lemma. But it splits also for $\mathfrak{p} \notin S$ because of the choice of S. Thus $\mathcal{A} \otimes K$

splits everywhere locally, and hence $\mathcal{A} \otimes K$ splits by Theorem 8, applied to $\mathcal{A} \otimes K$. Thus the cyclic field K is a splitting field for $\mathcal{A}$, i.e., $\mathcal{A}$ is cyclic.

The story about the proof of Lemma 10 has been told in the book [Roq05a]. A generalization to arbitrary fields has been given in [LR03].

The next problem is to prove Artin's Reciprocity Law, i.e., the product formula

$$(2.15) \qquad \prod_{\mathfrak{p}} \left(\frac{a, K}{\mathfrak{p}} \right) = 1$$

for the norm symbols defined locally as in (2.13). Hasse succeeded by comparing the abelian field $K|k$ with a suitable cyclotomic field, similar to but not the same as Artin's crossing method. Here the algebras do not come into play in an essential way.

We see that in the year 1932 Hasse and Noether (together with Artin and other collaborators) had almost finished her project of creating a new foundation of class field theory, built on the idea of first clearing up the local situation and then combining Local to Global. When I say "almost" then we have in mind that Furtwängler's norm theorem from the year 1912 still could not be explained by the new "abstract" algebraic methods. Furtwängler had made heavy use of classic analytic methods. In principle Hasse was not against the use of analysis in justifying class field theory; see, e.g., his letter to Noether of November 19, 1935. But when he says there that no one had yet succeeded in building class field theory without analysis then he seems to have ignored the new ideas of Herbrand and Chevalley of which he was certainly already informed, although perhaps not in their final form.

2.6 Herbrand and Chevalley

Jacques Herbrand was a brilliant young mathematician from Paris who visited Germany as a Rockefeller student in the academic year 1930/31. He studied with Von Neumann in Berlin, with Artin in Hamburg and with Emmy Noether in Göttingen. On the recommendation of Noether he participated in Hasse's workshop in Marburg on class field theory, February 26 to March 1, 1931. (See her letter of February 8, 1931.) This workshop had been planned as part of the project of developing the new class field theory. In particular Hasse was interested in a proof of the Norm Theorem; at that time this had not yet been established. As a consequence of Herbrand's meeting with Hasse there resulted a lively correspondence between them – until Herbrand's mortal accident in August 1931.

In a letter of May 27, 1931 Herbrand wrote to Hasse: *"I am glad to be able to inform you about a 'reasonable' proof of Takagi's theorems ... "*. And in his next letter of June 29, 1931: *"Since my last letter I have found great simplifications for class field theory ... "*. Those simplifications were indeed great; it induced Artin to write to Hasse: June 16, 1931: *"I am excited about the new tremendous simplifications of class field theory by Herbrand and Chevalley."* Let us briefly indicate the nature of those simplifications:

Jacques Herbrand 1931

Consider a cyclic extension $K|k$ of number fields, with Galois group G. Consider the norm class group $k^\times/N_{K|k}(K^\times)$, which today can be written as the cohomology group $H^0(G, K^\times)$, where G operates on the multiplicative group $K^\times$. We write briefly $H^0(K)$. Let J_K be the so-called idele group of K, defined as a subgroup of the direct product $\prod_{\mathfrak{p}} K_{\mathfrak{p}}^\times$ consisting of those elements which are $\mathfrak{p}$-adic units at almost all primes. (The infinite primes are included.) $K^\times$ is diagonally embedded into J_K, let $C_K := J_K/K^\times$ be the idele class group. Thus we have an exact sequence

$$1 \longrightarrow K^\times \longrightarrow J_K \longrightarrow C_K \longrightarrow 1$$

In today's language of cohomology this leads to an exact diagram of cohomology groups (note that G is cyclic):

$$
\begin{array}{ccccc}
H^0(K^\times) & \longrightarrow & H^0(J_K) & \longrightarrow & H^0(C_K) \\
\uparrow & & & & \downarrow \\
H^1(C_K) & \longleftarrow & H^1(J_K) = 1 & \longleftarrow & H^1(K^\times) = 1
\end{array}
$$

We have inserted $H^1(K) = 1$; this is Hilbert's "Theorem 90" from his "*Zahlbericht*". Similarly $H^1(J_K) = 1$. The norm theorem for cyclic extensions claims that the map $H^0(K) \to H^0(J_K)$ is injective. From the exact sequence it is immediate that this is equivalent to $H^1(C_K) = 1$. This statement is called the "Principle Genus Theorem" (*Hauptgeschlechtssatz*).[15] This theorem had been part of Takagi's construction of class field theory; a proof was given in Hasse's class field report [Has27a]. In her letter of July 21, 1932 Noether conjectured that the Principal Genus Theorem is much more elementary then the Local-Global Principle. This was precisely the idea of Herbrand who gave a simple direct proof of the Principle Genus Theorem, using his now well-known "Herbrand's Lemma". (Compare [Roq14].) These are the "tremendous simplifications" which Artin mentioned in his letter to Hasse. And, due to the diagram above, we see that $H^1(C_K) = 0$ immediately gives the Norm Theorem, without having to cite Furtwängler's old results of 1912.

---

[15]That name has historic roots, going back to Gauss.

Emmy Noether wrote a paper where she gave a proof of the Principal Genus Theorem that is also valid for non-cyclic Galois fields [Noe33a]. But in her proof she relies on the Local-Global Principle. Herbrand's new idea is to reverse the argument: first proving the Principal Genus Theorem which then immediately leads to the Local-Global Principle.

Going back to the exact diagram above, we see that the map $H^0(J_k) \to H^0(C_K)$ is surjective. From this we can deduce Artin's Reciprocity Law in a straightforward way, at least in the cyclic case. For abelian extensions Chevalley's methods, which he had used in the local case, apply.

Concluding, we see that Noether's idea of using non-commutative algebras for clearing up the structures of class field theory has been successful. But at the same time we see that precisely this success was responsible for the fact that her pets, the algebras, were increasingly replaced by their skeletons, the factor sets. A central simple algebra can be described by a factor set of one of its Galois splitting fields, which in the cyclic case is just an element of the base field modulo norms. The rules for handling these factor sets had to be established and finally led to general cohomology theory – which in turn replaced Noether's theory of algebras as a tool for understanding class field theory. The seminar notes by Artin and Tate in Princeton 1951/52 [AT68] mark the summit of this development.

We do not know whether Emmy Noether had fully realized this development which started in the early 1930s. Let me cite a remark by Hasse, a witness of those times 1920–1940, at the end of his sketch of the history of class field theory in [CF67]:

> *The sharply profiled lines and individual features of the magnificent edifice [class field theory] seem to me to have lost somewhat of their original splendor and plasticity by the penetration of class field theory with cohomological concepts and methods, which set in so powerfully. . .*

Would Noether have reacted similarly? After all, the cohomological development still followed the guiding principle of her work, as formulated by Van der Waerden in [vdW35]:

> *All relations between numbers, functions and operators only become transparent, generalizable and truly fertile when they are separated from their particular objects* [here: the algebras] *and are reduced to conceptual relationships* [here: the factor systems].

We see that Noether's work on class field theory in the 1920/30s, together with that of Hasse, Artin and others, prepared the way for the cohomological development which in turn reflected her own predilection for abstraction.

Chapter 3

The Letters

Contents

© Springer Nature Switzerland AG 2022
P. Roquette, F. Lemmermeyer, *The Hasse - Noether Correspondence
1925 - 1935*, History of Mathematics Subseries 2317,
https://doi.org/10.1007/978-3-031-12880-6_3

3.1   Postcard of January 19, 1925

Unique factorization.

Sehr geehrter Herr Hasse!

With your question about the principal prime ideal decomposition you probably mean, with a view to Hensel, that this decomposition should be satisfied for *every ideal* of the ring, not only for *every element*, i.e. every principal ideal.[1] The second condition is much weaker, as seen in the example of the polynomial ring in several variables where every element can be decomposed uniquely into prime elements, whereas for ideals every case which is possible under the ascending chain condition can arise.[2]

In the first case it follows from the given condition that *every ideal is principal*, and additionally the existence of a unit element, if one still considers ideals of the form $(n\mathfrak{a})$ — where n is a natural number — as decomposed.

Conversely, these assumptions also directly give the desired decomposition — and I do not believe that one can phrase this axiomatically in a different way. For the existence of the ideal basis implies the ascending chain condition, so every ideal becomes a product of finitely many indecomposable ones, i.e., ones that cannot be written as the product of two proper divisors. These indecomposables are principal prime ideals, however, since it follows from the assumptions that divisibility implies a product decomposition, from $\mathfrak{a} \equiv 0 \pmod{\mathfrak{b}}$ we obtain $\mathfrak{a} = \mathfrak{b}\mathfrak{c}$ with a proper divisor $\mathfrak{c}$, when $\mathfrak{b}$ is not the unit ideal.

More sensible than the above is a direct axiomatization of the Hensel rings, by which I mean for fixed π the system of all π-adic numbers of non-negative order.[3] They satisfy the Innsbruck conditions,[4] because the ideal theory of algebraic integers applies to them — ring without zero divisors, with unit, double chain condition for every ideal distinct from zero,[5] algebraically integrally closed in the field of fractions —, and additionally the further condition of having *only one* prime ideal distinct from the zero and the unit ideal. If conversely a ring satisfies these conditions, then every ideal distinct from the zero and the unit ideal is of the form $\mathfrak{p}^r$; if now $\pi \equiv 0 \pmod{\mathfrak{p}}$, but $\pi \not\equiv 0 \pmod{\mathfrak{p}^2}$, then the principal ideal generated by π is necessarily equal to $\mathfrak{p}$; therefore it is a principal ideal ring, and even a Hensel ring. — At the same time, this remark gives the reason why every ideal in a number field modulo a fixed ideal becomes principal; for if we represent the quotient ring modulo a fixed ideal as a direct sum of primary ones, then there are again only ideals of the form $\mathfrak{p}^r$, i.e., principal ones. But direct sums of "generalized" cyclic groups belonging to distinct primary ideals are again cyclic. (One can, as Krull[6] has recently found, obtain the Hensel rings simply by extending the ring of integers to a new ring by forming quotients by all numbers not divisible by a fixed prime ideal.)

I have not yet inserted the citation of Krull's paper into your article,[7] which you will probably do during the proofs.

Mit besten Grüßen, Ihre Emmy Noether.

Notes

[1]The present postcard apparently is the answer to an enquiry by Hasse which is not known to us. But we can deduce from Noether's answer that Hasse had asked for an axiomatization of those rings in which the theorem of principal prime ideal decomposition holds. Noether immediately notes that this question can be interpreted in two ways: Either every ideal, or every principal ideal of the ring should be uniquely decomposable as a product of principal prime ideals. The first interpretation leads to principal ideal rings, the second to unique factorization rings. Here, Noether discusses the first case although Hasse seems to be mainly interested in the second, as the correspondence in the next letters will show.

As far as we know, previously Hasse had not shown particular interest in axiomatization. But apparently he was impressed by Emmy Noether's talk at the DMV-meeting in Innsbruck in September 1924 which he attended, and now he started to concern himself with such questions. (DMV = *"Deutsche Mathematiker Vereinigung".*)

In Innsbruck, Noether had presented the axiomatic characterization of what is today called the class of "Dedekind rings", i.e. those rings in which every ideal decomposes uniquely as a product of prime ideals. This class in particular contains the rings of integers of algebraic number fields. See [Noe24]. Apparently Hasse is now trying to axiomatically characterize the unique factorization rings.

It is not clear what Noether has in her mind when she referred to Hensel in this context. Perhaps Hasse had mentioned his name in his enquiry. In any case Noether seems to relate this to what she calls "Hensel π-adic rings", that is the completions of the rings of algebraic integers of an algebraic number field. These are discrete valuation rings, for which Noether gives an ideal-theoretic characterization in this letter. But as said above already, Hasse is interested here not so much in the principal ideal domains or discrete valuation rings, but in the unique factorization domains. Compare also the letter from 11 December 1926.

[2]It is not quite clear what Noether means when saying that "every possible case can arise". One possible interpretation is that in polynomial rings of several variables, which are unique factorization domains, there are ideals whose minimal number of generators becomes arbitrarily large — provided the number of variables is suffiently large.

[3]The following remarks imply that Noether here means the valuation ring of a discrete valuation in a given field. She hence gives a ring-theoretic description of discrete valuation rings. — She writes π for a prime element (uniformiser) of the valuation.

[4]"Innsbruck conditions": Those are Noether's well-known axioms for a Dedekind ring. As said above already, Noether had talked about those at the Innsbruck DMV meeting in 1924. But her major paper *"Abstrakter Aufbau der Idealtheorie"* (Abstract development of ideal theory) [Noe24] on this subject appeared the following year only; perhaps this is the reason why she mentions these axioms explicitly here, so that Hasse would understand her remarks. — Also compare the letter from 3 November 1926.

[5]That is, the ascending chain condition for ideals, and the descending chain condition for the those ideals which contain some fixed non-zero ideal.

[6]At the time of the present postcard, Wolfgang Krull was *Privatdozent* in Freiburg. As a student he had been in Göttingen in 1920/21, where he had joined Emmy Noether's circle. Since then he kept in close mathematical contact with her. The strong influence of Emmy Noether on Krull's mathematical work is apparent. Noether's ideal theory found entry in Krull's book about ideal theory [Kru35].

[7]This is Hasse's article "Two existence theorems about algebraic number fields", which appeared 1925 in the *Mathematische Annalen* [Has25]. Emmy Noether worked as an unofficial editor for the *Mathematische Annalen*. The paper had been dated by the author (i.e. Hasse) as of 15 December 1924, and carries the date of receipt of 16 December 1924. At the time of the present postcard the paper had already been accepted for publication, and Hasse should insert only one citation, namely to Krull's paper [Kru24]. Hasse did so during proof-reading; the citation appears in the footnote on page 1 of his article.

This paper of Hasse's (or more precisely, the fact that Noether knew this paper) would be the start of the co-operation between Hasse and Noether in the area of algebras. See Noether's postcard from 4 October 1927.

3.2 Letter of November 03, 1926

Discriminants, Symmetric functions, Axiom V.

Lieber Herr Hasse!

I am very pleased to hear that you have read my discriminant paper[8] with so much interest and pleasure; many thanks also for the – very important – insertion "and operations".[9] For now it is Grell's[10] intention to concern himself with the theory of ramification;[11] with his theory of norms in arbitrary orders he is well-prepared for this;[12] I suspect that here the length of the composition series will take the place of the exponent. However, it does not seem easy to find the truly natural concepts; once one has those, the theory will presumably be more transparent than what is currently known for the maximal order.[13]

Now to your questions! Concerning the symmetric functions — the proof is due to Furtwängler, by the way, and Van der Waerden merely related it to me — you have forgotten to replace $\overline{a}$ by a. It is as follows: When $S(x_1, \ldots, x_n)$ is symmetric, we get $S(x_1, \ldots, x_{n-1}, 0) = G(\overline{a}_1, \ldots, \overline{a}_{n-1})$. I form the *symmetric* function $H(x_1, \ldots, x_n) = S(x_1, \ldots, x_n) - G(\overline{a}_1, \ldots, \overline{a}_{n-1})$; hence $H(x_1, \ldots, x_{n-1}, 0) = 0$ since $a_i(x_1, \ldots, x_{n-1}, 0) = \overline{a}_i$; hence divisible by x_n and therefore by a_n, so we have achieved the desired reduction of the degree.[14]

Regarding Axiom V.[15] *Assumptions and notation:* Let R be a ring satisfying the five axioms, K its field of fractions; $\mathfrak{o}$ the unit ideal and $\mathfrak{p}$ an arbitrary ideal in R. Further let $\mathfrak{o} : \mathfrak{p}$ be the quotient, *taken in K*, of the two R-modules (ideals) $\mathfrak{o}$ and $\mathfrak{p}$, that is all those elements σ of K such that $\mathfrak{p}\sigma \equiv 0 \pmod{\mathfrak{o}}$, i.e. *integral* in R. We therefore have, when we consider everything as R-modules in K: $\mathfrak{p} \cdot (\mathfrak{o} : \mathfrak{p}) \equiv 0 \pmod{\mathfrak{o}}$ and further $\mathfrak{p} \equiv 0 \pmod{\mathfrak{p} \cdot (\mathfrak{o} : \mathfrak{p})}$; the second claim follows from $\mathfrak{o} \equiv 0 \pmod{\mathfrak{o} : \mathfrak{p}}$ by multiplication with $\mathfrak{p}$. Since $\mathfrak{p}$ has no proper divisor in R except the unit ideal (axioms I, II, III), we have — since $\mathfrak{p} \cdot (\mathfrak{o} : \mathfrak{p})$ is in R — only the two options: $\mathfrak{p} \cdot (\mathfrak{o} : \mathfrak{p}) = \mathfrak{o}$ or $\mathfrak{p} \cdot (\mathfrak{o} : \mathfrak{p}) = \mathfrak{p}$. We have to show:

1) $\mathfrak{p} \cdot (\mathfrak{o} : \mathfrak{p}) = \mathfrak{o}$ implies $\mathfrak{q} = \mathfrak{p}^\rho$.[16]

2) $\mathfrak{p} \cdot (\mathfrak{o} : \mathfrak{p}) = \mathfrak{p}$ is impossible because of integral closedness.

Concerning (1) From $\mathfrak{q} \equiv 0 \pmod{\mathfrak{p}}$ it follows that

$$\mathfrak{q} \cdot (\mathfrak{o} : \mathfrak{p}) \equiv 0 \pmod{\mathfrak{p} \cdot (\mathfrak{o} : \mathfrak{p})} \equiv 0 \pmod{\mathfrak{o}},$$

hence an integral ideal.

Here the exponent[17] has decreased by precisely one; for $\mathfrak{p}^\rho \equiv 0 \pmod{\mathfrak{q}}$ implies $\mathfrak{p}^\rho \cdot (\mathfrak{o} : \mathfrak{p}) = \mathfrak{p}^{\rho-1} \equiv 0 \pmod{\mathfrak{q} \cdot (\mathfrak{o} : \mathfrak{p})}$ and conversely. We therefore have $\mathfrak{q} \cdot (\mathfrak{o} : \mathfrak{p})^\rho = \mathfrak{o}$, since with a positive exponent we have divisibility by $\mathfrak{p}$ and hence an integral ideal.[18]

Put differently: $\mathfrak{o} : \mathfrak{p}$ is the inverse element $\mathfrak{p}^{-1}$ to $\mathfrak{p}$ in the sense of the multiplicative group formed by the ideals.

Concerning (2) We have to show: (2_a) Every element of $\mathfrak{o} : \mathfrak{p}$ is integral over R; (2_b) $(\mathfrak{o} : \mathfrak{p})$ contains elements not in R; hence R cannot be integrally closed under assumption (2).

(2_a)): Usual determinant method: Let $p_1, \ldots, p_k$ be an ideal basis (R-module basis) of $\mathfrak{p}$, and σ an arbitrary element of $\mathfrak{o} : \mathfrak{p}$; then the assumption gives:

$$\sigma p_1 = a_{11}p_1 + \cdots + a_{1k}p_k;$$

$$\vdots$$

$$\sigma p_k = a_{k1}p_1 + \cdots + a_{kk}p_k,$$

hence, since R is without zero divisors and the $a_{\mu\nu}$ are in R:

$$\begin{vmatrix} a_{11} - \sigma & & a_{\mu\nu} \\ & \ddots & \\ a_{\mu\nu} & & a_{kk} - \sigma \end{vmatrix} = 0; \quad \text{integral dependence.}$$

(2_b) Let p be arbitrary in $\mathfrak{p}$; then for the principal ideal $\mathfrak{o}p$ generated by p, axioms I, II, III give the decomposition into primary components $\mathfrak{o}p = \mathfrak{q} \cdot \mathfrak{a}$, where $\mathfrak{a}$ denotes the product of those not belonging to $\mathfrak{p}$. Let $\mathfrak{q}$ be of exponent ρ (since $p \equiv 0 \pmod{\mathfrak{p}}$ certainly $\rho > 0$), so $\mathfrak{p}^{\rho-1}\mathfrak{a} \not\equiv 0 \pmod{\mathfrak{o}p}$ and therefore there exists an element t of $\mathfrak{p}^{\rho-1}\mathfrak{a}$ for which $t \not\equiv 0 \pmod{\mathfrak{o}p}$ holds.

This means, however, that t/p is not in R; but $t \equiv 0 \pmod{\mathfrak{p}^{\rho-1}\mathfrak{a}}$ implies $t\mathfrak{p} \equiv 0 \pmod{\mathfrak{o}p}$ or $t/p \equiv 0 \pmod{\mathfrak{o} : \mathfrak{p}}$, so we have constructed the desired element.

Here everything is truly reduced to the concept of integral closedness, and is completely natural. For the converse[19] Krull passes to the ring of fractions $R_{\mathfrak{p}}$, which is obtained by adding all elements prime to $\mathfrak{p}$ to the denominator. This must become a principal ideal ring, since there are now only primary ideals belonging to $\mathfrak{p}$, and no ideal between $\mathfrak{p}$ and $\mathfrak{p}^2$ (compare to my discriminant paper, top of p. 102.).[20] As a principal ideal ring without zero divisors $R_{\mathfrak{p}}$ is integrally closed; since this is true for *every* $\mathfrak{p}$, it is also true for R. It seems to me that one really does need my deductions with the abelian group here; that the integral and fractional ideals form an abelian group is proved by $\mathfrak{o} : \mathfrak{p} = \mathfrak{p}^{-1}$.

Krull has not gone on to write to me about the passage from all $R_{\mathfrak{p}}$ to R. But one has to show that in *every* representation $\sigma = a/b$ of an element *not* in R at least *one fixed* $\mathfrak{p}$ occurs in the denominator; that σ can hence not be in the respective $R_{\mathfrak{p}}$. (To put it better: if σ is not in R, then there is at least one $\mathfrak{p}$ such that σ is not in $R_{\mathfrak{p}}$ (already holds under axioms I to IV, but is more complicated to prove; Krull builds his generalized $\mathfrak{p}$-adic numbers on this). Excuse this improvised ending!) At least also the converse is shorter![21]

Beste Grüße, Ihre Emmy Noether.

For the improvised ending! One simply proves the proposition as follows:

Assumption: R satisfies axioms I to IV; $R_\mathfrak{p}$ consists of those and only those quotients $\sigma = a/b$, where a and b are in R, for which $b \not\equiv 0 \pmod{\mathfrak{p}}$.

Claim: *R is the intersection of all* $R_\mathfrak{p}$.[22]

R is contained in the intersection of all $R_\mathfrak{p}$; therefore it only remains to show: if σ is not in R, then there is an $R_\mathfrak{p}$ not containing σ; i.e. a $\mathfrak{p}$ such that *every denominator* of σ is divisible by $\mathfrak{p}$. (All $R_\mathfrak{p}$ contained in the field of fractions).

Let therefore $\sigma = a/b = c/d$ and a not divisible in R by b. But ad does become divisible by b in R. Under the primary components of $b = \mathfrak{q}_1 \cdots \mathfrak{q}_t$ there must be at least *one*, say $\mathfrak{q}$, such that $a \not\equiv 0$ $(\mathfrak{q})$; since otherwise a would be divisible by b — the least common multiple. But $ad \equiv 0$ $(\mathfrak{q})$ and $a \not\equiv 0$ $(\mathfrak{q})$ imply $d \equiv 0$ $(\mathfrak{p})$ by definition of the primary ideal. Hence $\mathfrak{p}$ must occur in every denominator; σ is not in $R_\mathfrak{p}$.

Hence also under axioms I to IV: An element is integral if and only if it is integral at all finite "places".

Beste Grüße, Ihre Emmy Noether.

[*The following text is in Hasse's handwriting on the back of the last sheet.*]

Detailed proof to p. 3 of the accompanying letter. *Claim*: $\mathfrak{q} = \mathfrak{p}^\rho$.

By deduction at bottom of page 2, $\mathfrak{q}(\mathfrak{o} : \mathfrak{p})$ is integral, $\mathfrak{p}$-primary, hence either $= \mathfrak{o}$ and then $\mathfrak{q} = \mathfrak{p}$ $(\rho = 1)$ or $\mathfrak{p} \mid \mathfrak{q}(\mathfrak{o} : \mathfrak{p})$; then by deduction at bottom of page 2 $\mathfrak{q}(\mathfrak{o} : \mathfrak{p})^2$ integral, $\mathfrak{p}$-primary, hence either $= \mathfrak{o}$, then $\mathfrak{q} = \mathfrak{p}^2$ or $\mathfrak{p} \mid \mathfrak{q}(\mathfrak{o} : \mathfrak{p})^2$; then $q(\mathfrak{o} : \mathfrak{p})^3$

integral... Since we cannot have $\mathfrak{q}(\mathfrak{o} : \mathfrak{p})^{\rho+1}$ integral for arbitrarily high ρ, since otherwise $\mathfrak{p}^{\rho+1} \mid \mathfrak{q}$ would follow, yet $\mathfrak{q} \mid \mathfrak{p}^{\rho}$, after finitely many steps $\mathfrak{q} = \mathfrak{p}^{\rho}$ must follow.

Notes

[8]Emmy Noether had submitted her manuscript of the discriminant paper [Noe27] to *Crelle's Journal* in March 1926. Hasse was already on the editorial board of *Crelle's Journal* at that time and apparently had carefully read Noether's paper (as he used to do with every submitted paper). The article can be considered as a sequel to her major work [Noe26a], in which she had axiomatized Dedekind rings. (Noether's axioms for these are listed below in this letter, but were already mentioned in her preceding postcard from 19 January 1925.) In the article discussed here, Noether considers a Dedekind ring R and an R-order T in a separable field extension; in this situation, the prime divisors of the (relative) discriminant of $T|R$ are characterized – generalizing the classical theorem due to Dedekind for the maximal orders in number fields.

[9]Noether had written that a ring was characterized by its *Gleichheits-beziehungen* (equality relations); Hasse had substituted *Gleichheits- und Verknüpfungsbeziehungen* (equality relations and operations). (See lines 4 and 5 on page 85 of [Noe27].) This was not trivial at that time, since the abstract concept of an axiomatically defined ring was not yet widely known; it had been introduced only a few years before by Emmy Noether in her article about rings with maximum condition, i.e., Noetherian rings [Noe21].

[10]Heinrich Grell belonged to the closer circle of students of Emmy Noether. At the annual meeting of the DMV in Prague 1930 he gave a talk about the ramification theory of orders [Gre30b], but he did not fully publish the details of the results presented there; only much later the *Mathematische Zeitschrift* published an article by him on the ramification theory of orders [Gre36].

[11]In the introduction to her paper Emmy Noether remarks that her discriminant theorem "can only be the first theorem in a general ramification theory of orders", and she compares the situation with Hilbert's ramification theory in number fields, where everything depends on the ramification exponents of the prime ideals. In this letter, Noether repeats the conjecture from her paper that in the general theory of ramification the length takes the place of the exponent. – Today we would consider the order T as a (possibly singular) "arithmetic curve", where the theory of ramification is subsumed in the concept of "classification of singularities".

[12]Compare [Gre30a].

[13]We see here an explicit formulation (for one example) of Noether's maxim which penetrates her entire work, namely (as Van der Waerden phrased it in his obituary for Emmy Noether): *"All relations between numbers, functions and operators only become transparent, generalizable and truly fertile when they are separated from their particular objects and are reduced to conceptual relationships."*

[14]This concerns a certain detail in the proof of the fundamental theorem of symmetric functions. Hasse was then occupied with writing the second volume of his text book "Höhere Algebra" (Higher algebra) [Has27c]. (The first volume had already appeared [Has26c].) In this second volume Galois theory was to be covered, for the first time in a textbook in the context of abstract field theory à la Steinitz [Ste10]. Probably Hasse intended to present Galois theory using the fundamental theorem of symmetric functions, as was then widely done. Emmy Noether had related a proof to him, which she had heard from Van der Waerden. Hasse had now apparently sent her an elaboration of this proof, and in the letter she makes some critical comments. At the same time she informs him that the proof is not due to Van der Waerden, but due to Furtwängler. Note that the $\overline{a_i}$ denote independent variables (today it is usual to write capital letters A_i instead).

As we can tell from the correspondence Hasse-Furtwängler, Hasse subsequently asked Furtwängler for permission to use this proof in his book. Furtwängler answered in a letter from 9 November 1926: *"Sehr geehrter Herr Hasse! I am happy for you to use this proof in your book. As I also cover some other things in a different way than is usually done, I shall briefly sketch how I deal with this section in my algebra lectures..."* However, in the published version of Hasse's book he had changed his mind and he expressly avoided the fundamental theorem of symmetric functions. Nevertheless he formulated the fundamental theorem, but for its proof Hasse refers to the forthcoming volume 3 of his book. (Yet a third volume never appeared. But Hasse explicitly mentions that the forthcoming proof in volume 3 will be due to Furtwängler, as Emmy Noether relates in this letter.) Compare also the notes to the following postcard from 10 November 1926.

[15]In the rest of the letter Noether explains to Hasse certain details in her theory of Dedekind rings. She had presented her theory at the DMV-meeting 1924 in Innsbruck, and Hasse had been in the audience. At the time of this letter, the notes of her lecture had already appeared in print [Noe26b]. But in the meantime Krull had been able to simplify some of her proofs. It seems that Hasse had heard about this and had asked Noether about details since Krull's paper [Kru28] was not yet published. Here we see Noether's answer.

In her lecture Noether had characterized Dedekind rings R by five axioms I–V. The first four axioms I–IV express in today's terminology that the following conditions (i) and (ii) are satisfied for R; the last axiom V requires condition (iii).

(i) R is a Noetherian integral domain.

(ii) Every prime ideal $\mathfrak{p} \neq 0$ of R is maximal.

(iii) R is integrally closed.

Here Noether gives Krull's simplified proof that under conditions (i) and (ii), condition (iii) is equivalent to the following:

(iv) Every primary ideal $\mathfrak{q}$ is a power of the associated prime ideal $\mathfrak{p}$.

(This indeed easily implies that every ideal factors as a unique product of powers of prime ideals.)

Noether's elaborations in this letter are essentially the same as Krull's in [Kru28], except that Krull immediately passes to the localization $R_{\mathfrak{p}}$ with respect to a prime ideal $\mathfrak{p}$, whereas Noether at first does not use this localization method. Only at the end Noether mentions the ring of fractions and refers to Krull for this. Apparently the passage from the ring to a localization was not yet as familiar as it is nowadays. This can also be inferred from the fact that Noether had asked her student Heinrich Grell to systematically present the ideal-theoretic situation under localization; compare [Gre27a].

In this context, the following remark of Van der Waerden's seems interesting, which mirrors the situation back then. In [vdW75] he writes about the sources of his book "Modern Algebra" where he had covered Noether's theory of Dedekind rings in the second volume 1931). In [vdW75] he writes: *"Emmy Noether's proofs were simplified, making use of ideas of Krull contained in §3 of Krull's paper [Kru28]. Emmy Noether was a referee for his paper, and she told Artin about it. Artin simplified Krull's proof and presented it in a seminar in Hamburg, in which I participated. Artin's simplified proof was reproduced in §100."* (He refers to §100 of the first edition of "Modern Algebra".)

We therefore find in this letter an intermediate stage of the proof, between Krull's and Artin's simplification, the latter of which then found entry into Van der Waerden's book.

[16]Here $\mathfrak{q}$ is an arbitrary $\mathfrak{p}$-primary ideal.

[17]Since $\mathfrak{q}$ is assumed to be $\mathfrak{p}$-primary, there is a $\rho \geq 0$ such that $\mathfrak{p}^{\rho} \equiv 0 \bmod \mathfrak{q}$. Noether calls the smallest such ρ the "exponent" of $\mathfrak{q}$.

[18]Marginal note in Hasse's handwriting: "See details on the back of the accompanying sheet." We give Hasse's explanations at the end of this letter.

[19]The "converse", which Noether proves here, says the following: *If every ideal $\neq 0$ of an integral domain R can be uniquely represented as a product of powers of prime ideals, then R is integrally closed.*

[20]At the top of page 102 of the discriminant paper [Noe27], the following theorem can be found: *If a multiplication ring $\mathfrak{H}$ has only one prime ideal distinct from the zero and the unit ideal, then $\mathfrak{H}$ is a principal ideal domain.* Here Noether uses "multiplication ring" to refer to an integral domain in which the non-zero ideals from a group under ideal multiplication, i.e., a Dedekind ring. Incidentally, Noether had already mentioned this theorem in the previous postcard from 19 January 1925; apparently she does not remember.

[21]Noether means "shorter than in my paper [Noe26a]".

[22]Today we know that this is true for *every* integral domain R and its localizations $R_{\mathfrak{p}}$, where $\mathfrak{p}$ ranges over the maximal ideals of R.

3.3 Postcard of November 10, 1926

Primitive elements.

Lieber Herr Hasse!

I also wish to give you Galois' proof for the existence of the primitive element, for possible inclusion in your book.[23] It appears in Galois (in his major treatise on solutions in radicals) for the case of the adjunction of several roots of the same equation, which is essentially the same thing.

Assumption: The field K contains infinitely many elements; $K(a,b)$ is separable;[24] to show: $K(a,b) = K(c)$.

Let $f(x)$ and $g(x)$ be the polynomials, irreducible over K, with zeroes a and b, which — in suitable field extensions — split into distinct linear factors. Let therefore $a = a_1, a_2, \ldots, a_n$ be the distinct zeroes of $f(x)$, $b = b_1, b_2, \ldots, b_m$ those of $g(x)$. Let s in K be chosen such that the $a_i + sb_k$ are all distinct.

But this means: if I put $j = a + sb$, then the product $F(x) = (j-(a+sx))(j-(a_1+sx))\cdots(j-(a_n+sx))$ vanishes only for the one zero $x = b$ of $g(x)$. Now $F(x) = f(j - sx)$ is a polynomial over $K(j)$ which shares with $g(x)$ in $K(j)$ only the one linear factor $(x - b)$; thus b is in $K(j)$ and therefore so is a.

I think that this proof, which uses *no* theorem about the representation by symmetric functions, is much more beautiful than the usual one due to Lagrange! As far as I can see — I have not thought it through completely — the theorem of symmetric functions is therefore not used anywhere in Galois theory.

Concerning the deduction $\mathfrak{q} = \mathfrak{p}^\rho$:[25] I had summarized this in the remark: the exponent of $\mathfrak{q} \cdot (\mathfrak{o} : \mathfrak{p})$ has decreased by precisely one; hence $\mathfrak{q} \cdot (\mathfrak{o} : \mathfrak{p})^\rho$ will be of exponent zero and integral, therefore equal to $\mathfrak{o}$. (For $\mathfrak{p}^\rho \equiv 0 \pmod{\mathfrak{q}}$ implies $\mathfrak{p}^\rho \cdot (\mathfrak{o} : \mathfrak{p}) \equiv 0 \pmod{\mathfrak{q} \cdot (\mathfrak{o} : \mathfrak{p})}$ and $\mathfrak{p}^\sigma \equiv 0 \pmod{\mathfrak{q} \cdot (\mathfrak{o} : \mathfrak{p})}$ implies $\mathfrak{p}^{\sigma+1} \equiv 0 \pmod{\mathfrak{q}}$). It is essentially your conclusion.

Beste Grüße, Ihre Emmy Noether.

Notes

[23]This again concerns the second volume of Hasse's "Higher Algebra" [Has27c]. The following proof, which is due to Galois as Noether writes, was indeed included by Hasse in his book with the reference to Galois. This proof avoids the use of symmetric functions, which was discussed in the preceding letter from 3 November 1926. — It appears that Hasse sent this proof to Furtwängler, together with the message that he had now decided to cover the Primitive Element Theorem in his book with this method, in order to avoid the use of symmetric functions. In a postcard from 23 November 1926 Furtwängler answers: *"Sehr geehrter Herr Hasse, I am familiar with the proof of the theorem $K(\alpha, \beta) = K(\theta)$ which you sent me after a message from Emmy Noether, as I have always covered it in my lectures in this way, without using the theorem of symmetric functions. I proved the theorem of symmetric functions for its own sake, and for its use with more general resolvents."*

[24]Noether still uses Steinitz' terminology "extension of the first kind" from his major work "Algebraische Theorie der Körper" (Algebraic theory of fields) [Ste10], as was usual at the time. The term "separable" was introduced later only, by Van der Waerden in his textbook "Modern Algebra" [vdW31].

[25]This again concerns the consequences of Noether's fifth "Innsbruck axiom" which she had discussed in her previous letter. Apparently Hasse had written that he had to think more carefully about the conclusion $\mathfrak{q} = \mathfrak{p}^\rho$ (compare the preceding letter from 3 November), and this is Noether's answer.

3.4 Letter of November 17, 1926

Galois theory without symmetric Functions. H's class field report.

Lieber Herr Hasse!

I have realized that indeed in my last lectures I did *not* need the theorem of symmetric functions,[26] but then your Theorems I and II come only *after* the fundamental theorem; normal is defined as "identical with all its conjugates"; which agrees with your definition due to Theorem II. I want to sketch this for you, since the fundamental theorem is reached very quickly in this way. I do not know, however, whether this requires too many changes in your manuscript.[27]

Definition: 1) Two finite (field) extensions L_1 and L_2 of K are called *equivalent* with respect to K, if they can be mapped isomorphically in such a way that for K the identity map is obtained. 2) Two finite extensions L_1 and L_2 which are equivalent with respect to K are called *conjugate* (with respect to K) if there exists a common extension field of L_1 and L_2. 3) A finite extension L of K is called *normal* if it is identical to all its conjugates. The system of automorphisms of L which yield the identity for K is the Galois group.

Theorem of the primitive element: If L is a finite *separable* extension of K, then L is simple, $L = K(z)$ (Galois proof).

Main theorem. If L is normal and finite over K, then the degree $[L : K]$ is equal to the order of the Galois group G.

For if $L = K(z)$, then the group is given by $(z \sim z)$, $(z \sim z')$, $\ldots$; since z must pass to a conjugate root under every automorphism, and the same z' corresponds to the same automorphism.

First half of the fundamental theorem. If M is an intermediate field of K and L, and H the largest subgroup which leaves M invariant, then M consists of the totality of all elements invariant under H. For L is normal over M, and H is equal to the group of L/M. If M' were an extension of M which also admits H, then the degree $[L : M]$ would become equal to $[L : M']$ by the main theorem; since both groups would be equal to H. Hence $M' = M$.

Second half of the fundamental theorem. If H is a subgroup, M the fixed field of H, then H consists of the totality of automorphisms which M admits. For the product $\prod_H t - z$ admits only H with indeterminate t; therefore also with suitable specialization of t in K (here we assume infinitely many elements in K again, as in the proof of the primitive element).

Both together give the fundamental theorem. (Always assuming a separable extension!)

Consequences I. If β is arbitrary in L, then the product $(x - \beta)(x - \beta') \cdots (x - \beta^{(v)})$ over all *distinct* values of β is in K. For *this* admits the *whole* group, as the decomposition into cosets shows.

II. The field $K(\alpha_1, \ldots, \alpha_r)$ is normal, if $\alpha_1, \ldots, \alpha_r$ are all the roots of an irreducible equation. For every isomorphism induces in the conjugate field only a permutation of $\alpha_1, \ldots, \alpha_r$ (because of the common overfield); hence all conjugates are identical.

III. In a normal field, every $g(x)$ irreducible in K splits completely if it has a linear factor $(x-\beta)$. For $g(x)$ is divisible by the product of all distinct $(x - \beta)(x - \beta') \cdots$, which arise through the group, since $g(x)$ permits the group. This product is in K according to I, and is therefore — as $g(x)$ was assumed to be irreducible in K — equal to $g(x)$.

IV. If this property of complete splitting is satisfied for every irreducible $g(x)$, then L is normal. For this is then in particular satisfied for the equation of the primitive element; therefore we have $L = K(z) = K(z, z', z'', \dots)$, thus normal by II.

I am glad that you continue to bring order into Takagi.[28] I notice more and more how your report simplifies the exposition; one only has to compare with Hilbert's one; how much effort is required there which is unnecessary *today*! [29]

Beste Grüße, Ihre Emmy Noether.

Notes

[26]This concerns once again the presentation of Galois theory in the second volume of Hasse's "Höhere Algebra" [Has27d]. In the preceding postcard from 10 November 1926, one week past, Noether had already suspected that in Galois theory the theorem of symmetric functions is not needed at all, but she had not been sure of it. Here we have her confirmation that it is indeed unnecessary. — It is not clear which lectures Emmy Noether means when she says that she could make do without symmetric functions. There is a list of the lectures which Noether gave over the course of the years in Göttingen. (We are indebted to Dr. Koreuber for kindly making this list, which she compiled, available to us.) According to this, in the winter term 1923/24 Noether had announced a problem session (Übungen) with "Talks on field theory". Presumably these are the lectures in question, since in the same term, on 27 November 1923, Emmy Noether gave a talk together with R. Hölzer on "Galois Theory in arbitrary fields" for the Göttingen Mathematical Society. A summary of this talk can be found in the annual report of the DMV, volume 33 (1924), p. 119 (italic). Apparently Galois theory was presented, already at that time, on the basis of Steinitz' abstract field theory.

[27]Hasse did *not completely* include this treatment of Galois theory in his book. He did indeed give the definition of normal extensions as suggested by Noether. Nevertheless, he placed Noether's consequences I—IV *before* the Main Theorem and thus *before* the Fundamental Theorem of Galois Theory (in Noether's terminology).

A few years later, the book [Ste30] appeared with a reprint of Steinitz' fundamental article [Ste10]. The book was edited by Hasse and his assistant Reinhold Baer, and they added comments and additions. In the current context the *appendix* of that book is relevant, in which the editors give a presentation of Galois theory since this had not been covered by Steinitz. The development of Galois theory in this appendix is mostly along the lines which Noether sketches in the present letter. At one point of this appendix they say explicitly that symmetric functions are avoided. Emmy Noether's influence is apparent.

[28] At that time Hasse was working on the continuation [Has27a] of the first part of his Class Field Report. Apparently Noether had already studied the first part where class field theory according to Takagi had been systematically presented, but without complete proofs. In [Has27a] Hasse would give full proofs. Hasse also planned a second part devoted to explicit reciprocity laws. But that second part was not completed until 1930 [Has30a] since Hasse rewrote that part after Artin had proved his general reciprocity law in 1927.

[29] Apparently the style of Hilbert's *Zahlbericht* (Report on Numbers) [Hil97] was not sufficiently abstract in Emmy Noether's view. When Noether compares Hilbert's Zahlbericht with Hasse's "*Klassenkörperbericht*" [Has26a] then she does not mean that Hasse's report should *replace* Hilbert's. Hasse had designed his report as a *sequel* to Hilbert's. In a letter from Hilbert to Hasse on 5 November 1926, Hilbert expresses his wish that Hasse writes the report in such a way that everyone who is familiar with his (Hilbert's) *Zahlbericht* might understand it "comparatively easily". Hasse makes this precise in the preface to his report by stating that he assumes Chapters I–VII of Hilbert's report as known. — According to Olga Taussky [TT81], in later years Noether was still rather critical about Hilbert's *Zahlbericht*. She cited Artin's opinion that the report delayed the development of algebraic number theory by decades. We do not know Artin's motivation for his criticism.

3.5 Letter of December 11, 1926

Class number 1. Unique factorization. Class field theory axioms?

Lieber Herr Hasse!

Among your results it seems of general interest that you now have a simple norm criterion for absolute class number one in a number field; and I think you should mention this already in the introduction.[30]

Besides, I personally am so strongly focused on ideals that I believe that the truly characteristic properties of integral domains can be stated in terms of ideals and not as properties of elements; and the fact that your conditions really concern principal ideals seems to justify this.[31] It is therefore of interest to me whether Hensel can say something essential about the polynomial ring,[32] where polynomials can only give the algebraic objects of highest dimension — whereas the others, such as space curves, necessarily lead to ideals. (On page 10, by the way, concerning the integral domain with unique factorization of elements as coefficient ring, you do not have to restrict yourself to *one* indeterminate; although more indeterminates with a distinguished one of them mean the same, which is what you probably wanted to say with your restriction.) When you pass from the polynomial ring to the functional domain[33] — i.e. adjoin all quotients of functions primitive in u as units (§3, 4. in my ideal theory[34]) you recover the statement of Theorem 2; [35] and furthermore finite rank with respect to the Abelian group of these units. This rank becomes the sum of exponents of a basis element; (e.g.:

1. Polynomial ring of x and y, coefficients from field; modulo x^2 the element y and its powers, generally every polynomial not divisible by x, become units; for let $g(x, y) \not\equiv 0 \pmod{x}$; then $\frac{g(x,y)}{g(x,y)+ux^2} \equiv 1 \pmod{x^2}$; denominator primitive; hence $1, x$ are a complete residue system, linear independent modulo units.

2. Integral polynomial ring in x, principal ideal domain $2 \cdot 3^2$; the residue ring becomes the direct sum of residue rings by 2 and 3^2; mod 2 every element becomes a unit, the residue ring a field, hence rank *one*. mod 3^2 we have rank 2 with respect to the units, i.e. with respect to the residue field mod 2; and so on generally.)

With this, the polynomial domain is essentially covered; but I also see that your function χ is constructible in any case: we only have to put $\chi(\alpha) = 2^{\text{Sum of exponents}}$, when $\alpha = \pi_1^{e_1} \ldots \pi_k^{e_k}$

therefore $\chi(\alpha) = 2^{e_1 + \cdots + e_k}$. Then (1) to (4) follow in the situation of Theorem 2; and in the situation of Theorem 1 when you pass to the functional domain. (Obviously $\chi(0) = 0$ by definition; $\chi(y) = 1$, which corresponds to exponent zero of all prime elements). Of course the norm criterion for the number field is nevertheless more sensible! But your note will become shorter and more general. The passage to the functional domain — for the necessity of the criterion — is of course necessary if and only if the double chain condition is not satisfied mod every $\mathfrak{m} \neq (0)$; (4) is then only satisfied in the functional domain.

Theorem (9) is indeed not quite correct, and will presumably be changed anyway.[36] In §3, 1.[37] I have the following hypotheses which are sufficient for transfer of the chain conditions (and probably also essentially necessary; at least they are *all* used in the given course of proof): In $\mathfrak{I}$ the 5 axioms are satisfied, including *integral closedness* (this is not assumed in your version, but is true if Theorem (1) or (2) holds); $\mathfrak{I}$ is an order; the extension is separable; then the chain conditions transfer; in particular $\overline{\mathfrak{I}}$ has a finite module basis with respect to $\mathfrak{I}$: $\alpha_1, \ldots, \alpha_k$; and therefore the finite rank transfers as in §3, 2. (The correction hence already has to be put *before* the question mark.) Therefore the remark about $\overline{\chi}(a) = \chi(N(a))$ holds; and the proof proceeds directly if Theorem 2 holds in $\mathfrak{I}$, since the $\alpha_1, \ldots, \alpha_k$ are linearly independent in this case. If only axioms I to V are satisfied, one has to force linear independence by passing to appropriate localizations (point rings); this has been done by Grell much more generally.

You write "simple" extension; this could also be inseparable. In this situation — and using my results — Artin and Van der Waerden have shown the validity of the transfer, but under the hypothesis that the root ring of $\mathfrak{I}$ is finite with respect to $\mathfrak{I}$.[38] This note has recently been published in the Göttinger Nachrichten, I do not know if Van der Waerden has already sent it to you — together with an application given by me to the finiteness of invariants.[39]

The matters of the norm still have to be discussed here further (instead of requiring $\mathfrak{I}$ to be integrally closed, one can also require that in the induced integrally closed domain of $\mathfrak{J}$-integral elements of the quotient field the double chain condition holds; but this is almost the same.)

It would be very nice if one could look closely into class field theory by way of axioms! Where have the things by F. K. Schmidt appeared; he has not yet sent them here! [40]

Beste Grüße, Ihre Emmy Noether.

I am sending back the paper, unsubmitted, so that it will not lie around here until Monday.

The question arises whether it is not better to change (2) to $\chi(\alpha\beta) = \chi(\alpha) + \chi(\beta)$ and alter (1) accordingly; and then introduce the *sum of exponents itself*. This is what we call the length of the ideal; namely the length of a composition series from the unit ideal to (α); where we possibly have to pass to the functional domain. This arises as a very important invariant in Grell's investigations[41] (compare § 10[42]).

Notes

[30]Apparently Hasse had sent to Noether a draft manuscript of his article on unique factorization into prime elements [Has28]. (Compare the postcard from January 19, 1925.) He had not sent it for publishing in the *Mathematische Annalen* for which Noether acted as an (unofficial) editor. He only wished to inform her about the content and to know her opinion. The paper later appeared in *Crelle's Journal*. Hasse's result implies a criterion for class number 1 in an algebraic number field K, namely: *For any two integers α, β in K with $N(\alpha) \geq N(\beta) > 0$ there exist two integers ν, μ such that*

$$0 < |N(\nu\alpha - \mu\beta)| < |N(\alpha)|$$

with ν prime to β. (N denotes the norm function.) This is more general than the euclidean algorithm. As Noether suggests in this letter, in the published version Hasse mentions this norm criterion in the introduction already.

Today this criterion is known under the name of "Dedekind-Hasse". The criterion was found unpublished in Dedekind's writings after his death, and was included by Noether in Dedekind's Collected Works [Ded31] (Volume 2, No 38) . There Emmy Noether remarks: *"The criterion given here has been found again very recently only — in the context of more general investigations."* And then she cites Hasse's paper [Has28]. We can deduce that Noether did not yet know Dedekind's manuscript when she wrote the present letter to Hasse, since otherwise she would surely have voiced her famous saying *"this is already in Dedekind"*. Only later, when she edited Dedekind's works (jointly with Robert Fricke and Öystein Ore), she found out that Hasse's criterion had already been known to Dedekind. The latter had found e.g., that in the field $\mathbb{Q}(\sqrt{19})$ there is no euclidean algorithm but the class number is 1 because of the validity of the above criterion.

More generally Hasse proves: An integral domain I is a principal ideal domain if and only if there exists a non-trivial multiplicative function $\chi: I \to \mathbb{N}$ which satisfies the condition above for χ instead of N.

[31]In his paper [Has28] Hasse specifically mentions that he had actually been looking for a criterion for rings with unique factorization into prime elements, a criterion which is not only sufficient but also necessary. But he had not found such a criterion. His criterion is necessary and sufficient for the smaller class of principal ideal domains only, as Noether immediately notes.

[32]Hensel had published a proof of the following theorem in *Crelle's Journal* [Hen27]: *If I is an integral domain with unique factorization into prime elements, then the same is true for the ring of polynomials $I[x]$*. This theorem might have been considered well-known at the time, as a simple consequence of Gauss' Lemma, but apparently did not exist in the early literature in this form for an abstract ring I. Hensel's method of proof consists of giving explicitly a suitable function χ on the polynomial ring $I[x]$, which satisfies a condition similar to Hasse's condition above.

The approaches of Hasse and Hensel are therefore quite similar, also in terms of notation. The date of Hensel's publication is after the date of the present letter but Hasse, being close to Hensel, knew Hensel's method. Indeed Hensel points to Hasse's paper as a first step of "refining" his method.

Later Krull established a necessary and sufficient condition for an integral domain to be a unique factorization domain by modifying Hasse's axiom for the function χ suitably [Kru31]. (Compare letter of 02 November 1930.) See also the paper of F. K. Schmidt [Sch28], where he presented a criterion a la Noether, i.e., with ideal theoretic methods. Since F. K. Schmidt was in close contact with Hasse, we can safely assume that Hasse had encouraged him to look for such criterion, according to the proposal of Emmy Noether. F. K. Schmidt's result has been superseded by Krull's [Kru31].

[33]The term "functional domain" as Noether defines it was standard at the time. In the present case it refers to the localization of the polynomial ring $R[u]$ with the primitive polynomials as admissible denominators, i.e. those polynomials whose coefficients have greatest common divisor 1 (under the assumption that R has unique prime factorization). This functional domain is a principal ideal domain. Today such rings are known under the name

"Kronecker function rings" and are fairly well studied. (Compare for instance [FL05].) — In the published version [Has28], Hasse mentions in a footnote that functional domains had ben pointed out to him in a *"letter from Miss E. Noether"*. Obviously he refers to the present letter.

[34]Noether refers to her paper "Abstrakter Aufbau der Idealtheorie in algebraischen Zahl- und Funktionenkörpern" [Noe26a] in the *Mathematische Annalen*.

[35]The theorem numbers refer to Hasse's manuscript.

[36]There is no "Theorem 9" in Hasse's publication [Has28], it seems to have been removed. From Noether's discussion we may infer that "Theorem 9" concerned the passage to the integral closure in a finite separable field extension.

[37]Here Noether refers to her paper [Noe26a] again.

[38]The "root ring" here is $\mathfrak{J}^{1/p}$, where p is the characteristic.

[39]Compare [Noe26b] and [AvdW26], and also [Gre35].

[40]At that time Hasse was in correspondence with F. K. Schmidt, who on Hasse's suggestion had started to develop class field theory for algebraic function fields in one variable with finite field of constants. Apparently Hasse had the idea to put class field theory on an axiomatic basis and thereby to treat both number fields and function fields simultaneously, and he had told Noether about this. However, in this context F. K. Schmidt writes to Hasse in a letter of 6 December 1926: *"I had not intended the axiomatic foundation of class field theory mentioned by you in this generality..."* And later in the same letter: *"It would of course be very nice if one could phrase the finiteness hypotheses in such a way that they characterize all fields for which every relatively Abelian field extension is a class field; similarly to how all fields with usual ideal theory are characterized by descending and ascending chain conditions after Miss Noether..."* It seems that Noether had believed that F. K. Schmidt already had found an axiomatization of class field theory, which was not the case. — The paper after which Noether enquires had been submitted by F. K. Schmidt in November 1926 as a first preliminary announcement in the Erlanger Berichte [Sch27]. Compare [Roq01]. — For the idea of an axiomatization of class field theory compare also the following letter from 3 January 1927.

[41]Presumably Noether refers to the article [Gre27b], which had not yet appeared at the time of this letter.

[42]Section 10 refers to Noether's paper [Noe26a], where composition series are treated.

3.6 Letter of January 03, 1927

Global class field theory. Complex multiplication.

Göttingen, 3.1.1927

Lieber Herr Hasse!

I am very much interested in your ideas regarding class field theory. They go in the direction I have always thought of after Dedekind-Weber (Crelle 92).[43]

I. Formal part,

II. Abstract Riemann surface.

The formal part — which singles out an integral domain — becomes essentially ideal-theoretic; the hypotheses are my five axioms, from which one can — by passage to the localization — additionally recover your 6th axiom of the principal ideal property. Here the hypotheses are less restrictive than you assume; e.g. my discriminant paper belongs here and applies to functional domains, i.e. algebraic functions of *arbitrarily* many variables. The arbitrary choice of the integral domain does not exhaust the entire field, however. In contrast I do believe that you are right with your supposition — number field or algebraic functions in *one* variable — in the second part; but then one has to use *all* integral domains. I think the model for the subdivision in classes here has to be the Dedekind-Weber polygon classes (divisor classes), to which would correspond your classes of valuations. It would be very nice if you could build class field theory on this — instead of on ideal classes! However, one does have to add a finiteness condition; for in the situation of Dedekind-Weber (characteristic zero) there are infinitely many polygon classes.[44]

This gives you some more remarks floating in the air. Many more successes for 1927![45]

Your Crelle anniversary paper seems to finally give a reasonable introduction to complex multiplication![46]

Beste Grüße, Ihre Emmy Noether.

Notes

[43]Noether is referring to the paper *"Theorie der algebraischen Functionen einer Veränderlichen"* (Theory of algebraic functions of one variable) which appeared in 1882 [DW82].

[44]After Noether's preceding letter, Hasse had apparently communicated some more precise thoughts on how he imagined an axiomatization of class field theory. We do not know the details of these thoughts. It is interesting that he already at this time was thinking about a simultaneous treatment of number fields and function fields in one variable, i.e. (with a finite constant field) the theory of what is today called *global fields*. In fact, Hasse seems to be thinking about a *valuation-theoretic* foundation of class field theory.

[45]Emmy Noether's remarks are not entirely floating in the air. The situation of Dedekind-Weber can be abstractly described as a field with a given set of absolute values, i.e. a multi-valued field. When Noether speaks of a "finiteness condition" that must be added, this can be interpreted as stipulating that these absolute values be either discrete with finite residue field or archimedean. Hasse himself always explicitly stressed the role of the *product formula* for valuations in his number-theoretic development of arithmetic. (See for instance [Has26e].) If one adds this product formula as an axiom, then by Artin-Whaples [AW45] one is working with either an algebraic number field or a function field in one variable with finite constant field, i.e. with a global field. In a global field one has class field theory, as is well known.

This would give a rather indirect axiomatization of class field theory, but would probably match Noether's ideas which she voices in this letter. The disadvantage is that this would not capture *local* class field theory. It should be noted, however, that local class field theory was not known at the time; it was only later discovered by Hasse [Has30c] and will play a role in later letters of Noether.

Later Artin-Tate[AT68] and also Neukirch [Neu86] gave axioms of a different kind for class field theory.

[46]This is Hasse's paper *"Neue Begründung der komplexen Multiplikation I. Einordnung in die allgemeine Klassenkörpertheorie"* (A new basis for com-

plex multiplication I. Presentation in terms of general class field theory) [Has26d], which had appeared in 1926 in the 100 years anniversary volume of *Crelle's Journal*. A second part appeared later, in which complex multiplication was developed without using class field theory [Has31b].

3.7 Postcard of October 04, 1927

Large splitting fields?

Göttingen, 4.10.27

Lieber Herr Hasse,

Could you tell me whether the general existence theorems about abelian fields[47] directly imply the following: for every n there exists at least one field (probably arbitrarily many) cyclic over the field of rational numbers of degree 2^n, such that its subfield of degree 2^{n-1} is real,[48] and such that (-1) can be represented in it as a sum of at most three squares (squares of fractional numbers). In the case of fields of degree 4 one can directly read this off from the parametric representation (e.g. Weber, kl. Lehrb. Section 93[49]); but in general I cannot get through formally.

This concerns an example in the theory of representations by matrices, namely the question of whether the degrees of the smallest fields in which an irreducible representation is possible are bounded. Here I call a field a smallest one (with respect to the representation) if this representation is not possible in any of its subfields.[50]

In the case of the quaternion fields, all number fields in which (-1) can be represented as a sum of at most three squares give irreducible representations by matrices. The conjecture above hence means that the degree of the smallest fields is not bounded. R. Brauer made (in Kissingen) the conjecture of unboundedness. His examples were more complicated than quaternion fields, however. It would follow that one knows much less about these small-est fields than I thought for a while.

Mit besten Grüßen, Ihre Emmy Noether.

Notes

[47] It appears that Noether refers to the existence theorems in Hasse's recent paper [Has26f].

[48] This condition of reality is always satisfied and hence could have been omitted. This is an example of Noether's habit of writing her postcards without checking the text one more time (in the same manner in which emails are often sent today). Otherwise she would surely have noticed that the reality condition is superfluous, as she notes herself in her next postcard from 19 October 1927.

[49] This refers to Heinrich Weber's *"Kleines Lehrbuch der Algebra"* (Little Textbook of Algebra) [Web12], as opposed to his better-known *"Lehrbuch der Algebra"* (Textbook of Algebra) in three volumes.

[50] At the 1925 DMV meeting in Danzig Emmy Noether had pointed out that the theory of matrix representations fits in the structure theory of algebras [Noe26c]. Afterwards she further developed this approach which would later culminate in the famous work [Noe26c], called *"one of the pillars of modern linear algebra"* (Curtis). At the time of the present postcard Noether's paper had not yet appeared, and Noether was concerned with the clarification of certain details, especially the properties of splitting fields. Let A be a central simple algebra of dimension n^2 over a field K. (Here, "central" means that the base K is the center of A. In Noether's time one said "normal algebra" instead of "central algebra".) An extension field L of K is called a *splitting field* of A if $A \otimes_K L$ is a full matrix algebra over L. This means that A has an irreducible matrix representation over L. In a letter to Richard Brauer on 28 March 1927 Noether had explained the principles of her algebra-theoretic conception of representation theory. But erroneously she claimed that every minimal splitting field of a central division algebra had degree n. At the DMV meeting in Bad Kissingen in September 1927 Brauer showed her a counterexample. Noether then asked whether at least the degrees of the smallest splitting fields are bounded. Also here Brauer produced a counterexample, as we can read in her letter. Noether thought this was too complicated, however, and she now wanted to know whether there are counterexamples for the usual quaternion algebra $\mathbb{H}$ over $\mathbb{Q}$. A field L over $\mathbb{Q}$ is a splitting field of $\mathbb{H}$ if and only if there is a non-zero element of $\mathbb{H} \otimes_{\mathbb{Q}} L$ whose norm vanishes, i.e., in

L there is a relation $a^2 + b^2 + c^2 + d^2 = 0$ with not all terms vanishing. If $a \neq 0$, say, then division by a^2 yields a representation of -1 as a sum of three squares; this explains Noether's question about fields with this property. — In his reply from 6 October 1927 Hasse proved the existence of such fields.

3.8 Hasse's Letter of October 06, 1927

Solution to Noether's question.

Halle, 6.10.27

Liebe Fräulein Noether!

Your conjecture is true, even if it is not a *direct* consequence of my earlier existence theorems. Nevertheless I can prove it with very similar methods:

I. Let the field to be constructed, of degree 2^n, cyclic over the rational field R, be called k; its subfield of degree 2^{n-1}, which was required to be real, is to be called k'. I construct k as *the* subfield of degree 2^n of a cyclotomic field k_p of p-th roots of unity. To ensure that k itself becomes imaginary, as the further demand that $-1 = \xi_1^2 + \xi_2^2 + \xi_3^2$ in k necessitates, $p - 1$ has to be divisible by *exactly* 2^n. In this case it is apparent that k is always imaginary and k' real, the latter because k' is contained in the subfield k_p' of k_p of degree $\frac{p-1}{2}$. Hence one first has to determine a prime number p such that

$$(3.1) \qquad p \equiv 1 \bmod 2^n, \quad p \not\equiv 1 \bmod 2^{n+1}$$

holds.

II. It follows from my paper Crelle 153, Satz 14 (page 128) that the equation

$$-1 = \xi_1^2 + \xi_2^2 + \xi_3^2$$

is solvable in k if and only if degree f or order e of the prime divisors $\mathfrak{l}$ of 2 in k is even.[51]

Since with the construction in I. the order $e = 1$, as 2 is not contained in the discriminant of k_p and hence not in that of k, we therefore have to ensure by choice of p that f becomes even. Since k_p over k is of odd relative degree $\frac{p-1}{2^n}$, the degree f_p of the prime divisors $\mathfrak{l}_p$ of 2 in k_p is an odd multiple of f and therefore even or odd as f is. Therefore it only comes down to ensuring that f_p is even. This f_p is the smallest exponent for which $2^{f_p} \equiv 1 \bmod p$. As one sees without difficulty by representing 2 mod p by a primitive root mod p, f_p is (by (1)) even if and only if 2 is not a 2^n-th power residue mod p. (One can also argue more directly in the following way: k is the class field for the group of 2^n-th power residues mod p over R, therefore $f > 1$, i.e. even as a divisor of 2^n, if and only if 2 is not a member of this group — decomposition law for the class field). Besides (1) we hence have to satisfy:

(3.2) $x^{2^n} - 2 \equiv 0 \bmod p$ unsolvable (in R)

III. Let us assume $p \equiv 1 \bmod 2^n$, but not necessarily $p \not\equiv 1 \bmod 2^{n+1}$. Then (2) is either unsolvable or already has 2^n non-congruent solutions. In the former case p has no prime factor of degree 1 in $R(\sqrt[2^n]{2})$; in the latter case p has precisely 2^n distinct prime factors of degree 1 in $R(\sqrt[2^n]{2})$ — in short: p becomes "completely split" in $R(\sqrt[2^n]{2})$. This last condition is further equivalent to p being completely split in the associated Galois field $R(\zeta_{2^n}, \sqrt[2^n]{2})$, where ζ_{2^n} is a primitive 2^n-th root of unity. Hence Condition (2) can be written as follows:

(3.3) p is not completely split in $R(\zeta_{2^n}, \sqrt[2^n]{2})$,

under the hypothesis $p \equiv 1 \bmod 2^n$. This last hypothesis means by the theory of cyclotomic fields that p is completely split in the cyclotomic field $R(\zeta_{2^n})$. Condition (1) demands in addition:

(3.4) p is not completely split in $R(\zeta_{2^{n+1}}) = R(\zeta_{2^n}, \sqrt{\zeta_{2^n}})$.

It therefore remains to prove the existence of such p, completely split in $R(\zeta_{2^n})$, which are completely split in *neither* $R(\zeta_{2^n}, \sqrt[2^n]{2})$ *nor* in $R(\zeta_{2^{n+1}})$, i.e. the existence of prime ideals $\mathfrak{p}$ of degree 1 in $R(\zeta_{2^n})$ for which both is the case. This again comes down to showing that those $\mathfrak{p}$ which are completely split in *either* $R(\zeta_{2^n}, \sqrt[2^n]{2})$ *or* in $R(\zeta_{2^{n+1}})$ together only constitute a proper fraction of all $\mathfrak{p}$. Now let r_n denote the relative degree of $\sqrt[2^n]{2}$ over $R(\zeta_{2^n})$, 2 is the relative degree of $\sqrt{\zeta_{2^n}}$ over $R(\zeta_{2^n})$, and let $\overline{r}_n$ denote the relative degree of the compositum ($\sqrt[2^n]{2}, \sqrt{\zeta_{2^n}}$) over $R(\zeta_{2^n})$. Then:

the $\mathfrak{p}$ completely split in $R(\zeta_{2^n}, \sqrt[2^n]{2})$ form precisely

the fraction $\dfrac{1}{r_n}$ of all $\mathfrak{p}$,

the $\mathfrak{p}$ completely split in $R(\zeta_{2^n}, \sqrt{\zeta_{2^n}})$ form precisely

the fraction $\dfrac{1}{2}$ of all $\mathfrak{p}$,

the $\mathfrak{p}$ completely split in $R(\zeta_{2^n}, \sqrt[2^n]{2})$ and $R(\zeta_{2^n}, \sqrt{\zeta_{2^n}})$ form

precisely the fraction $\dfrac{1}{\overline{r}_n}$ of all $\mathfrak{p}$,

and therefore the $\mathfrak{p}$ completely split in *either... or...* form precisely the fraction $\frac{1}{r_n} + \frac{1}{2} - \frac{1}{\overline{r}_n}$ of all $\mathfrak{p}$.[52] Hence one has to show:

(3.5)
$$\frac{1}{r_n} + \frac{1}{2} - \frac{1}{\overline{r}_n} < 1$$

IV. The relative degree r_n is not 2^n, but

$$r_n = 2^{n-1} \text{ for } n \geq 3,$$
$$r_n = 2^n \text{ for } n = 1, 2.$$

For $n = 1, 2$ this is clear, since $\sqrt{2}$ is an element of neither $R(\zeta_2) = R(-1) = R$ nor $R(\zeta_{2^2}) = R(i)$. For $n \geq 3$, however, $\sqrt{2}$ is an element of $R(\zeta_{2^n})$, so that certainly $r_n \leq 2^{n-1}$. But $\sqrt[2^2]{2}$ is not in $R(\zeta_{2^n})$. For if this were the case, then $\sqrt[2^2]{2}$ would be in the real subfield (of degree 2^{n-2}) of $R(\zeta_{2^n})$, therefore (because this is a Galois extension) also the quotient $i = \zeta_{2^2}$ of two conjugates to $\sqrt[2^2]{2}$, which is impossible. Therefore the claim about r_n is proven. The equation (5) therefore reduces to

$$\frac{1}{2^{n-1}} + \frac{1}{2} - \frac{1}{\overline{r}_n} < 1 \text{ for } n \geq 3,$$

$$\frac{1}{2^n} + \frac{1}{2} - \frac{1}{\overline{r}_n} < 1 \text{ for } n = 1, 2.$$

Apart from $n = 1$ this is already evident without using $\overline{r}_n$. For $n = 1$ we have $\overline{r}_n = 2^2$ as the relative degree of the compositum $(\sqrt[2^2]{2}, \sqrt{-1})$, which again gives the required relation.

Do excuse the copying pencil.[53] I use it to keep a copy for myself. Are you interested in publication? Then you could perhaps assemble a small note with your application to irreducible representations from the above, and give it to the Annals[54] or the *Göttinger Nachrichten*.

Mit besten Grüßen

Ihr H. Hasse.

Notes

[51]Here Hasse cites the paper [Has24], in which he proves the *local-global principle* for quadratic forms over number fields. The representability of -1 as a sum of three squares in a field K means, as Noether had written, that the quaternion algebra over K splits. Here for the first time we see explicitly the local-global principle for an algebra, namely the quaternion algebra. It would take more than four years for the discovery that this local-global principle holds for *every* central simple algebra over a number field. Compare Noether's letter from 10 November 1931.

[52]As Hasse gives no reference for the density theorem for completely split prime ideals used here, we can surmise that Emmy Noether knew it. Indeed it is an immediate consequence of the "analytic relations fundamental for class field theory", as Hasse puts it in his *Klassenkörperbericht* [Has30a] (Part

II, Chapter V. Section 24); here he means the behavior of L-series at the point $s = 1$. Today we view this density theorem as a special case of the general density theorem due to Chebotarev, which appeared 1925 in the *Mathematische Annalen* [Che26]. For the present case, however, already the weaker density theorem due to Frobenius from 1896 would suffice [Fro96]. — By the way, since Hasse relies on the density theorem he does not explicitly produce a field as demanded by Noether. He only showed the *existence* of such field.

[53] At that time letters were written by hand, usually with ink and a dip pen (or fountain pen). This was not suitable for creating a carbon copy, however. To this end, special "copying pencils" were available, with which one could create a second copy by applying sufficient pressure, but which were as permanent as ink. This letter is preserved because Hasse used such a pen to create a copy for himself.

[54] Hasse seems to have suggested the *Mathematische Annalen* because he knew that Emmy Noether worked as an unofficial editor there. "Unofficial" means that her name was not mentioned on the title page of the *Mathematische Annalen*. Nevertheless, those in the know wanting to publish a paper in Emmy Noether's area of interest in the *Mathematische Annalen* sent the manuscript directly to her in Göttingen. If Emmy Noether could recommend the paper, she forwarded it to Blumenthal, the editor in chief. The date of submission in the published article would be the day the manuscript was received by Emmy Noether. — Regarding Emmy Noether's work as an unofficial editor of the *Mathematische Annalen*, Hermann Weyl said in his eulogy at the memorial service in Bryn Mawr [Wey35]: *"Emmy Noether was a zealous collaborator in the editing of the* Mathematische Annalen. *That this work was never explicitly recognized may have caused her some pain."* — Hasse's article was not published in the *Mathematische Annalen*, however, but in the *Sitzungsberichte* of the academy in Berlin. See the following letter from 19 October 1927.

3.9 Postcard of October 19, 1927

Proposal for a joint paper with R. Brauer.

Lieber Herr Hasse!

Your proof has pleased me immensely; the matter hence does lie a bit deeper! I thought of a publication in the Berliner Sitzungsberichte, where up to now almost all short notices about representation theory are found. I have drafted a note of 5–6 pages, which I have sent to R. Brauer – Königsberg, so that he may add his part; it should then go under our names together, and also report on newer results. Your proof could then immediately follow as a short note; perhaps with the subtitle "from a letter to E. Noether", then no textual changes would be necessary at all! What should the main title be, and do you agree at all with my ideas? Should one bind the notes as one offprint or as two? [55]

I only have to wait for the response from R. Brauer. It would be possible after all, although not too probable, that he has proven the fact of unboundedness of the degrees of minimal splitting fields.[56] Then one would first have to see to what extent the example is still valuable.

By the way, I would now phrase my conjecture in your letter as follows: Cyclic field of degree 2^n over R, where -1 a sum of three squares. It is after all no additional demand that the subfield of degree 2^{n-1} is real; I had overlooked this fact earlier. I might send you the entire thing before it goes to Schur.[57]

Beste Grüße, Ihre Emmy Noether.

Notes

[55]The two notes of Brauer-Noether [BN27] and Hasse [Has27b] were bound together as one offprint.

[56]In fact, this was true. For minimal splitting fields of the quaternions, Brauer had been able to prove the unboundedness of the degrees directly, following the ideas of Hasse but without the use of class field theory. Moreover, in a footnote of [BN27] it says that meanwhile Brauer could show the existence of minimal splitting fields for *every even degree*, i.e., of every degree which is possible for the quaternions.

On first sight, these two notes of Brauer-Noether and of Hasse seem to deal with a relatively special and not too important topic, concerning the quaternions only. However, during the next years the same results were obtained for all central simple algebras over number fields. Thus these two small notes mark the beginning of a development of high importance, culminating in the so-called Hasse-Brauer-Noether theorem [BHN32] , which in turn was a big step towards our understanding of class field theory.

[57] Issai Schur, Berlin, was a member of the Prussian Academy of Sciences and therefore could submit to the Academy works for publication in the "Sitzungsberichte".

3.10   Postcard of October 26, 1927

Thanks for the new version. R. Brauer has more results.

Lieber Herr Hasse!

Many thanks for your manuscript; it has now indeed become clearer.[58] Do you not want to remark that using your Satz 13 it follows that (-1) can be represented as a sum of *two* squares as soon as it can be represented as a sum of *three*.[59] I had already wanted to write this recently, but forgot it, and am now getting back to it because R. Brauer calculated my quaternion criterion once more with his methods and there reached the representability of (-1) as a sum of *two* squares. Incidentally, this also holds for arbitrary totally imaginary fields, which may also contain parameters (or be of characteristic p and imperfect)![60] (It follows that in the prime field $\mathfrak{P}_p$ one can always represent (-1) as sum of two squares. Is this known?[61] There must therefore be representability criteria as a sum of squares!)[62] — I can insert the sentence; do you want to give me your phrasing, and whether it is to go in the text or as a remark?[63] — Brauer could prove in an elementary way that there are infinitely many n satisfying the conditions: for them there exists a p such that

$$(3.6) \qquad \begin{cases} p \equiv 1 \pmod{2^n} \quad p \not\equiv 1 \pmod{2^{n+1}}; \\ 2^r \equiv -1 \pmod{p} \text{ for at least one } r \end{cases}$$

He takes an arbitrary positive integer t; p a prime divisor of $2^{2^t}+1$; therefore $2^{2^{t+1}} \equiv 1 \pmod{p}$; let n be the biggest number such that 2^n divides $p - 1$; then $n > t$ and (1) is satisfied! Therefore there are infinitely many such n. — This is of course much less than your existence theorem, and your representability criterion is used. Nevertheless he can also verify that (1) implies the quaternion splitting criterion for the corresponding fields.

Beste Grüße, Ihre Emmy Noether.

It seems as if it will be another 14 days before I send the things to Schur because of the correspondence with Brauer. Do you want to see it before, or do corrections suffice for you?

Notes

[58] Apparently Hasse had sent a revised version of his letter of 6 October 1927 for publication.

[59] This refers to Satz 13 of Hasse's paper [Has23] in *Crelle's Journal*. In the previous letter Hasse had referred to Satz 14 from this paper. Satz 14 gives a criterion for the representability of an arbitrary element $\mu \neq 0$ of a number field K as a sum of 3 squares of the field. Satz 13 gives a criterion for the case $\mu = 0$, i.e. the representability of -1 as a sum of two squares. As Noether remarks, the criterion of Satz 14 for $\mu = -1$ implies that the criterion of Satz 13 is also satisfied. Both criteria refer to the behavior of the places 2 and ∞ in the field K.

[60] In the publication of Brauer and Noether [BN27] this is justified using the identity

$$(c^2 + d^2)(a^2 + b^2 + c^2 + d^2) = (ac + bd)^2 + (ad - bc)^2 + (c^2 + d^2)^2.$$

It is also said that this follows from the norm product formula for quaternions; compare Noether's remark on the next postcard from 1 November 1927 that she could *"only find something like this by way of quaternion fields or similar"*. There is a postcard from Noether to Richard Brauer from 2 November 1927, on which she explicitly calculates the formula above as a consequence of the norm product formula of the quaternions. She adds: *"Here the thing was a lot of fun!"* — Incidentally the identity above (beside two further identities which achieve the same) can already be found in Gauss' papers [Gau00].

[61] Yes, this was known. The theorem seems to have been first proved by Lagrange in the context of the four squares theorem. Hasse pointed this out

in his response to Noether; compare Noether's reaction in her next postcard from 1 November 1927.

[62]Here Noether mentions a general problem of algebra, which goes beyond the present context. Perhaps Noether discussed this question with Van der Waerden at some point, since he put the following question in the annual report of the DMV (volume 42) in 1932:

> If -1 is a sum of 3 squares in a field, then also of 2 squares; if of 5, 6 or 7, then also of 4; if of 15 or fewer, then also of 8.

In the following volume 43 of the annual report, several solutions can be found. One of these was submitted by Richard Brauer. He notes there that the first part of the question had already been covered in his paper with Emmy Noether [BN27], and that the solution for the other parts arose shortly thereafter. The solution of Van der Waerden's entire question was therefore known in Emmy Noether's circle. It is hard to imagine that Van der Waerden was not informed about this, since he also belonged to this circle and knew her papers. One possible explanation for his posing the question might be that he wanted to know whether the solution was known elsewhere. (This was not an uncommon motivation for a question in the annual report at the time.) Perhaps, inspired by Emmy Noether's question, he wanted to raise awareness of the general problem which necessarily follows from his question, namely the question after the minimal number of squares by which -1 can be represented, in an arbitrary field K. Today this minimal number is called the *level* of the field. Van der Waerden's problem suggests that the level of a field may be always a power of 2. This was only be solved much later by Pfister in [Pfi65]. Some years later Witt treated Pfister's theory in a new and very simple way, with the aid of his new concept of a "round" quadratic form. Witt's theory is covered in Lorenz's Lecture Notes [Lor70].

We see that this development was started by Noether's conviction stated in this letter: *"There must therefore be representability criteria as a sum of squares!"*

[63]This was now no longer necessary. In the published version [Has27b] Hasse only speaks of the representation of -1 as a sum of *two* squares and only refers to Satz 13, no longer to Satz 14.

3.11 Postcard of November 01, 1927

Local-global principle in higher dimension?

Lieber Herr Hasse!

Congratulations on the birth of your daughter, the Sunday child! Is there already a brother?

I had soon seen myself that the solvability of $-1 = x^2 + y^2$ in $\mathfrak{P}_p$ is trivial; I can only find such things, however, by way of quaternion fields or similar! [64]

Brauer's approach is sufficient for the intended application — unboundedness of the degrees;[65] your result, however, gives much more precise insight and is valuable most of all because one knows so little about the splitting fields of non-commutative fields. What we have in terms of general embedding theorems will go into the note without proof.

I believe that also in the case of higher transcendentals[66] one has enough ideal theory to begin to work on your theory of representability by quadratic forms. For — as already Kronecker essentially showed — the classical theory holds for ideals of highest dimension. You will find this explained most easily in my 5 axioms paper, Annalen 96, where I put this theory into context in §3 and 4. You can see that one has arithmetic in the usual sense in my discriminant paper,[67] where the functional domain is also covered; also in the paper of Ostrowski, Göttinger Nachrichten, cited there — Ostrowski there treats the theory of the different and so forth following Kronecker's definitions.[68] It would make me very glad if your theory could already be treated with these means; I do not think that one needs the much more complicated theory of ideals of lower dimension for your things. By way of the functional domains the variables — with one exception — are essentially moved into the units; it therefore has to be something like in the case of $\mathfrak{P}_p(x)$.[69]

Mit besten Grüßen, Ihre Emmy Noether.

Notes

[64]Apparently Hasse had written in his answer to Noether's question in the preceding letter that the representability of -1 as a sum of two squares in a finite field was indeed well-known. Noether here refers to Wedderburn's theorem that any simple algebra over a finite field splits, hence in particular also the quaternion algebra over the prime field $\mathfrak{P}_p$ with p elements.

[65]In the joint note [BN27] Brauer constructs minimal splitting fields for the quaternions with the help of an elementary lemma from number theory; he therefore does not need Hasse's local-global principle for ternary quadratic forms.

[66]Here Noether means function fields in several variables.

[67]The paper indicated is [Noe27]; compare also the letter from 3 November 1926.

[68]Compare [Ost19].

[69]Noether seems to believe that Hasse's local-global principle for quadratic forms might also hold in function fields of several variables, but she does not give a specification of the constant field (this should presumably be allowed to be either a number field or a finite field). Today we know that this is not the case in this form. Nevertheless the local-global principle for quadratic forms remains valid in function fields of one variable over a *finite* field — the prototype for which is the rational function field $\mathfrak{P}_p(x)$ over a prime field with p elements mentioned by Noether. This seems to have been explicitly verified in the doctoral dissertation of Hasse's PhD-student Rauter 1926 in Halle. Originally Hasse had suggested to Rauter to deal with arbitrary function fields of one variable over an arbitrary finite field. But Rauter apparently was not able to do this, and therefore his dissertation dealt with the rational function field over the prime field only. Rauter's dissertation was never published [Rau26]. Since he did not live in Halle, his contact with Hasse went through letters. These are preserved in the legacy of Hasse in Göttingen. Noether does not appear to know Rauter's dissertation. Later she would criticise a further paper of Rauter's [Rau28b]; compare the letter from 14 May 1928.

3.12   Postcard of December 26, 1927

Ordering offprints for joint notes. Ramification in orders.

Lieber Herr Hasse!

I am also sending you a proof copy of our note, which I would ask you to send back as I expect that you have also obtained a proof copy at the same time. The copy intended for the printer goes via R. Brauer to I. Schur, so that I can also send any wishes directly to Schur. You will be able to see already now whether you prefer binding together or joint sending.[70] For the offprints binding together poses no additional difficulties, as Schur wrote to me. To Brauer I suggested [that we order] 150–200 offprints, so that everyone [of us] has some for himself apart from the joint ones; I thought of 30–40 offprints for myself, for Göttingen and private distribution.

What number would you ask for? Und would you and Brauer arrange the distribution? I know my laxness (the discriminant paper is still half lying here) and your order in this regard. With the second proofs the printer should be informed about the distribution of the offprints. If we only publish together, you would have to be cited more precisely on p. 6, what is the precise title? [71]

You will be interested to know that I now also have approaches to ramification theory corresponding to the discriminant paper. I myself only want to treat the maximal order in a note in the Annalen,[72] and leave the elaboration for arbitrary orders to Grell, who had already thought about the questions. My definition of the different for $f'(x)$ — when $f(x) = (x - \alpha)(x - \beta) \ldots (x - \gamma)$ — comes down to: $f'(x) =$ trace of $f(x) : (x - \alpha)$. The place of $f(x)$ is taken generally by the relations between the ω.[73]

Best wishes for the new year and Christmas! How is your daughter?

Ihre Emmy Noether.

Notes

[70]The offprints of their two papers [BN27] and [Has27b] were indeed bound together and distributed in this form.

[71]The title of Hasse's note is "Existence of certain algebraic number fields". The title of the Brauer-Noether note is "On minimal splitting fields of irreducible representations".

[72]This *Annalen* paper was not published at all. Perhaps the approaches mentioned here finally went into the manuscript on the theory of the different which Noether gave to Grell for safe-keeping when she emigrated, and which was posthumously published in *Crelle's Journal*: [Noe50].

[73]In the posthumously published paper [Noe50], the symbols $\omega_1, \ldots, \omega_n$ denote an integral basis of an algebraic number field. These ω are meant here. The relations between the ω form an ideal in the integral polynomial ring of n indeterminates.

3.13 Postcard of January 06, 1928

Distribution of the offprints.

Lieber Herr Hasse!

I completely agree with your suggestion for distribution and will inform Brauer. He has proposed to order 200 offprints, these will suffice well for us all. Because of the binding together you must still write to Schur, however, also because of the mailing to the various places. My "private distribution" is essentially what you call a safety factor, and additionally distribution to a few good attendees of my lectures.

I am further interested that some should go to America, where much non-commutative work is done. I thought of: [74]

1. University of Princeton (New Jersey), Department of Mathematics; Lefschetz, Veblen, Wedderburn, Aleksandroff, H. Hopf (the latter ones as Rockefeller people there);

2. Chicago, Dickson

3. Vandiver in Univ. Texas, Austin

4. Wahlin, Univ. Missouri Columbia (These as referees.)

5. Olive Hazlett, Univ. Urbana, Illinois

6. MacDuffee, Univ. Columbus, Ohio

If you want to cover these as well, then 45 offprints would suffice for me including the Göttingen ones, otherwise I ask for 55. (Järnik, Prag and Suetuna, Tokyo are currently here). Skolem, Oslo, will be on your list? Also J. v. Neumann, Berlin.

Many thanks for your corrections;[75] I have made the formulations more precise and sent them to Schur.[76] The triviality of the [ ]-theorem[77] was stressed by an "offenbar"; I only put in the theorem because Schur always has the version with the conjugate representations. I have no remarks for your corrections; Schur's "elementarization" rather seems to be an "outmodification"![78]

I hope to get closer yet to higher arithmetic in the future; especially with representation theory; but it goes slowly.[79]

Beste Grüße, Ihre Emmy Noether.

Notes

[74]The following list, together with the list of names in the next postcard, is interesting because because it gives an indication with whom, outside of Göttingen, Emmy Noether was in scientific contact.

[75]It appears that Hasse had sent some corrections to the text of the Brauer-Noether paper.

[76]As already said, Issai Schur in Berlin was the editor of the *"Sitzungs-berichte"* of the Berlin Academy, where the notes of Brauer-Noether and of Hasse were to appear.

[77]The open brackets [ ] are in Noether's original. Presumably this means a passage of text in the proof copy, and Noether could assume that Hasse knew what this was about.

[78]Here, Noether refers to Hasse's corrections to his own paper. Schur had informed Hasse that he could also prove Hasse's existence theorem under discussion without ideal theory, with the aid of his supplementary theorems for the quadratic and biquadratic reciprocity theorem. At the end of his paper Hasse reproduces Schur's proof and compares it with his own proof. [Has27b]

[79]It is not clear what Noether means here by "higher arithmetic". Perhaps Noether refers to the ideal theory of the maximal orders of a simple algebra. Or she means class field theory. Or she means the connection between the two.

3.14 Postcard of May 02, 1928

Again: Distribution of offprints.

Lieber Herr Hasse!

I also have not received an invoice for offprints; it seems that the Reichsdruckerei has granted each of us three 50 free copies!

I still have 8 offprints left, but would like to keep them in reserve. It does not seem difficult to me to cut 5–10 people from the list; as there were a few who are not algebraically and arithmetically interested on it (for instance, I have no relation to Maier-Frankfurt; but as it occurs to me now, however, you probably do via the annual report problems[80]).

I do want to give you my list immediately: Noether, Bernays, Bernstein, Cohn-Vossen, Courant, Grandjot, Herglotz, Hilbert, Landau, Lewy, Neugebauer, Van der Waerden, Walther, Grell, Jarnik, Scorza, Wedderburn, Aleksandroff, Hopf, Stepanoff, O. Schmidt (Moskau), Chatelet, Tschebotarow (Chebotarev), Weber, A. Weil. The last two are students; A. Weil (from Paris) was here on a Rockefeller scholarship.[81] Chatelet is the only Frenchman who seriously thinks about these questions, also the number theoretic ones; he has now written a book about automorphisms of abelian groups, where a lot of things are done in a hypercomplex way. To him and Tschebotarow, who enquired after my things about characters of groups a few days ago, I still have to write. O. Schmidt (direct product) wrote a Russian book about group theory and representation; he was here last year, likewise Stepanov; subsequently he worked on Krull.[82] The Americans I have reduced a lot; I do still intend to write to Dickson. This is my somewhat colorful list! Suetuna will also go on it, who is here now.[83] I only returned on Sunday, so that you would not even have found me.

Beste Grüße, Ihre Emmy Noether.

Notes

[80]It appears that the mathematician Maier in Frankfurt was the editor of the problem-section in the *"Jahresberichte der DMV"* where sometimes Hasse (and other people around Noether) had given contributions. But we do not know.

[81]André Weil reports on his impressions from Göttingen, also about Emmy Noether, in his book [Wei93].

[82]The person "Krull" is not meant here, but the theorem of Krull on direct products, which is often called "Krull-Schmidt Theorem" today. The "Schmidt" named here is the Russian Otto Schmidt, who also appears on the list.

[83]This list, together with the list from the preceding postcard, seems interesting because it gives an indication with whom Emmy Noether was in scientific contact.

3.15   Letter of May 14, 1928

Inseparable finite extensions of function fields.

Lieber Herr Hasse!

You will see from the invoice above that together we have to pay for 100 offprints à 15 Pfennig = 15 Mark; I hereby send you my share of 5 M and would ask you to settle this bill together with yours.[84]

I have looked a bit at Rauter's[85] paper in the last Crelle issue; he forgot to say that he requires K to be a *separable* extension of k, although this is used several times, this is essential in some of the steps not given.[86]

By the way, ideal theory is still preserved in purely inseparable extensions, as F. K. Schmidt and in general cases Artin-Van der Waerden (*Erhaltung der Kettensätze (Preservation of the chain theorems)* — in the Göttinger Nachrichten 1926) have shown;[87] on the other hand, the theorems on different and discriminant change here. The fact that everything goes exactly like in the number field is based on the following assumptions, which are also the only ones used in the number field:

(1) K is a separable extension of k;

(2) the integral elements from k form a principal ideal domain;

(3) the residue ring with respect to every integral ideal from k (and therefore from K) consists of finitely many elements.

(1) and (2) give ideal theory, (3)[88] gives ramification theory; in particular the theory of inertia and ramification groups must agree,[89] as we are here concerned with a theory of the same residue rings,[90] whereas the subfields of K involved are given by Galois theory.

The real differences therefore *only* appear in the different behavior of the points at infinity.

Beste Grüße, Ihre Emmy Noether.

Notes

[84]Noether wrote this text on the margin and the back side of a bill which came from the printing shop, where the papers of Brauer-Noether and of Hasse had been printed for the Berlin Academy. It seems that the authors ordered more offprints than were allowed without charge; this bill covered

the (small) rest. Noether sent her part of the invoice amount to Hasse and asked him to settle the case; she did not like to deal with such formalities. But she included some text about mathematics which would be of interest to Hasse.

[85]Herbert Rauter was a teacher at a *gymnasium* in Tilsit (East Prussia) and earned a doctorate with Hasse in Halle in 1926. In his thesis he had transferred the results of Hasse's thesis to the case of a rational function field $\mathbb{F}_p(x)$ — namely the local-global principle for quadratic forms. The Crelle paper [Rau28b] now under discussion developed, among other things, the basics of arithmetic in algebraic function fields over finite fields, as well as Hilbert's ramification theory for Galois extensions of function fields.

[86]Noether's criticism led to a correction, which appeared in the same Crelle volume [Rau28a], and in which Rauter states that he indeed had to assume separability. (Noether still uses the terminology "extensions of the first and second kind" when the translation says "separable and purely inseparable"; compare the second note to the postcard of 10 November 1926).

[87]F. K. Schmidt had treated function fields in one variable in his paper [?] on analytic number theory in characteristic p. This paper appeared in 1931 but it was already finished in 1927 and had been used as a habilitation thesis; apparently Noether knew it. Artin and Van der Waerden treat extensions of such fields K of characteristic p for which the degree over the subfield K^p is finite. This is the case in function fields of one or more variables if the base field has this property. Compare [AvdW26]. Later Grell [Gre35] published the theorem entirely without additional assumption.

[88]Here the original contains a word which we could not decipher.

[89]Apparently this means that the theory of inertia and ramification groups in function fields agrees with the corresponding theory in number fields.

[90]This is not quite correct in case of wild ramification, as the residue rings in function fields (characteristic p) are not the same as the residue rings in number fields.

3.16   Postcard of August 12, 1929

Non-commutative ideal theory and different. Levitzky.

Lieber Herr Hasse! [91]

To avoid misunderstandings: Grell has *independently* achieved the transfer of the different theorem to the non-commutative.[92]

I had only given him the commutative proof back in the day, as I had no idea that he was interested in the non-commutative. I wrote immediately to Van der Waerden about the possibility of the transfer. Incidentally I was always interested in the "reduced" different (compare §24, 25 of the proof copy): $\mathfrak{d}_{\mathrm{red}}^{-1}$ consists of all μ with $\mathrm{Sp}_{\mathrm{red}}(\mu \mathfrak{o})$ integral.[93] One has to add additional considerations about reduced traces, about which I then — still without the right proof — wrote to Artin. In my lecture[94] about non-commutative arithmetic I have now given everything correctly: $\mathfrak{d}_{\mathrm{red}}$ is divisible by precisely those prime ideals which are contained at least quadratically in p. The non-reduced different, where the one-sided length is used instead of the exponent, is nevertheless also of interest: Grell should publish that one, as he has investigated the details more precisely. You will presumably also find it right that Grell's proof for the different in the commutative case should be published in a neutral place: it really is too amusing that three people thought of this proof. You appear to have been the first one (Artin and I in February).[95]

Artin's nice ideal theory[96] I have also presented non-commutatively: I had heard that this was possible through Artin and Van der Waerden. Much becomes easier than in Krull's proof.

Just one more question! Do you know of a vacancy as an assistant for an extraordinarily capable and likable person, who is however of Palestinian nationality; in Germany since 1922 and *nothing* about annoyingly Jewish. Levitzki graduated *summa* with a ring-theoretic foundation of Frobenian "relations between characters of groups and subgroups". He showed that the relations are *characteristic* for the subring class, going way further than Frobenius. Now he has done Galois theory of completely reduced rings.[97]

Beste Grüße, Ihre Emmy Noether.

Notes

[91]Since the last letter 15 months have passed. We remark that Noether spent the winter 1928/29 as guest professor in Moscow, on Alexandroff's

invitation, whom she knew from Göttingen. Noether returned to Göttingen at the end of May 1929.

[92]One month later, at the DMV meeting in September 1929 in Prague, Grell reported among other things about differents in the non-commutative situation. Compare [Gre30b]. He had submitted a more detailed version to *Crelle's Journal*. Given Noether's phrasing in the present letter, it appears that somehow the suspicion had arisen that Grell had obtained his proof from Emmy Noether; Noether strongly denies this here, but she admits that she had also thought this through herself and had written to Van der Waerden about it. Perhaps Hasse had received word from Van der Waerden. – Grell had been one of the first students of Noether in Göttingen. At the time of this letter he had a teaching position in Jena. Hasse had apparently taken a close interest in Grell's work on general ramification theory, and he had criticized his paper. In a postcard from 29 November 1929 to Hasse, Grell writes: "My ramification in general orders has lately taken such a form under the influence of your criticism that any improvement by myself seems impossible to me. . ." Grell had achieved his habilitation 1929 in Jena, and Hasse had been one of the referees. Grell's habilitation thesis was apparently not published. His paper concerning the ramification theory of orders appeared 1936 in the *Mathematische Zeitschrift*, but only for the commutative case [Gre36].

[93]The "proof copy" apparently refers to the proofs of Noether's article "Hypercomplex quantities and representation theory" [Noe29], which appeared in the *Mathematische Zeitschrift*. These are Van der Waerden's notes for Noether's lectures in the winter term 1927/28. Sections 24 and 25 of this paper contain the theory of the reduced trace and the reduced discriminant. However, only discriminants of algebras (hypercomplex systems) are considered there, and the only concern is the vanishing or non-vanishing of the discriminant, which indicates the existence or non-existence of the radical. In the present postcard, the interest is in the discriminants and differents of *orders* in algebras. — Compare also Noether's letter of 3 November 1926 and the postcard of 26 December 1927 for the commutative case.

[94]In the summer term of 1929, Emmy Noether gave a course on "Non-commutative arithmetic", Saturdays 11 AM-1 PM.

[95]In Artin's paper "On the arithmetic of hypercomplex numbers" [Art28a], different and discriminant of an order are defined and investigated. Artin mentions in the introduction, however, that there are also inessential divisors in his discriminant, and he says: *"I hope to return to a more useful definition of the discriminant on another occasion."* Artin did not publish anything more about this later, but it could be that he had thought of such a more useful definition in February 1929, possibly using the *reduced* trace, and that Noether refers to this. — Hasse also considered the different and discriminant

of maximal orders in his paper "About $\wp$-adic skew fields and their relevance for the arithmetic of hypercomplex number systems" [Has31d]; he reduces these to the local case and can therefore precisely determine the exponent to which the prime ideal divides the different. Perhaps Hasse had obtained these results already in 1929 and told Emmy Noether about them. Noether herself did not publish the discriminant theorems in the non-commutative situation. As mentioned in this postcard, she intended to leave this to her student Grell. In Deuring's report on algebras [Deu35a], in the discussion of differents and discriminants of maximal orders (Chapter VI, §5–6) he cited Noether's paper [Noe27] only, where the commutative case is treated. Probably Deuring knew that Noether had also obtained the theorem in the non-commutative case in the meantime, but he could not give a reference. (Or he did not want to, in order not to prevent Grell from publishing.) Deuring also cites Hasse's paper [Has31d] mentioned above.

[96]This refers to Artin's presentation of Van der Waerden's ideal theory of arbitrary integrally closed integral domains; this was taken up in the second volume of the textbook "Modern Algebra" [vdW31] under the name *"Quasigleichheit von Idealen"* (quasi-equality of ideals). We only know from this postcard that Emmy Noether had extended this to the non-commutative case.

[97]Levitzki's thesis appeared in [Lev31b] with an announcement in [Lev29]. For his Galois theory of semi-simple rings see [Lev31a].

3.17   Letter of October 02, 1929

Algebras and class field theory. First steps into cohomology.

Lieber Herr Hasse!

Since I first had to do Dedekind corrections[98] on my return,[99] I only get to answering your card today. Firstly — what I know about the connection between hypercomplex algebra and class field theory[100] is very modest and entirely formal; in fact, nothing except an interpretation of Hilbert's symbolic powers, for which I generally use the group ring as a domain of operators. For explanation I first want to assemble a few facts about the group ring, which can partly be found in the Zeitschrift article (the §s refer to this).[101]

1. If $a_1, \ldots, a_h$ are the elements of a finite group, I will denote by $\mathfrak{o}$ the corresponding *integral* group ring, i.e. the rational integral combinations of the a, taken as linearly independent, with the group multiplication rules.

   By $\mathfrak{S}$ I denote the associated hypercomplex system with respect to the field of *rational* numbers, i.e. all rational linear combinations of the a. Hence $\mathfrak{o}$ becomes an order in $\mathfrak{S}$ (not necessarily maximal, as the example of the quaternions shows; for cyclic groups of prime order $\mathfrak{o}e_1 + \mathfrak{o}e_2$ does become maximal (compare 2.)). $\mathfrak{S}$ becomes a ring without radical, therefore completely reducible (§26). $\mathfrak{S}$ shall be called the rational group ring. (In the paper *this* is denoted by $\mathfrak{o}$.)

2. The irreducible representations of groups and of the rational group ring are the same (§20, end); one therefore obtains all via direct sum decomposition. To obtain the absolutely irreducible ones the coefficient domain of $\mathfrak{S}$ must be extended algebraically; however, one can also proceed by first decomposing $\mathfrak{S}$ and only then perform the coefficient extension in the components. In the decomposition of $\mathfrak{S}$ itself the two-sided component of rank one corresponding to the identity representation splits off; and it is only this direct sum decomposition: $\mathfrak{S} = \mathfrak{S}e_1 + \mathfrak{S}e_2$, $e_1 + e_2 = 1$ (where we set $a_1 = 1$) and the orthogonality relations: $e_1 e_2 = e_2 e_1 = 0$; $e_1^2 = e_1$; $e_2^2 = e_2$; which are relevant for the moment. Here e_1 and e_2 are in the center of $\mathfrak{S}$.

3. If $\mathfrak{S}e_1$ gives the identity representation, then e_1 is given by $e_1 = \frac{a_1 + \cdots + a_h}{h}$ (remark by Levitzki), and from this we will obtain the connection with the norm. For we have: $a_i \cdot e_1 = e_1 \cdot a_i = e_1$ and therefore $e_1^2 = e_1$. However, there is only *one* idempotent which gives the identity map.

   The second component $\mathfrak{S}e_2$ becomes equal to the difference ideal

   $$\mathfrak{P} = \big((a_2 - 1), \ldots, (a_h - 1)\big)$$

   (a_1 equal to 1, therefore $a_1 - 1 = 0$). This follows from my general considerations regarding difference ideals, but can

of course also be verified directly, for instance in the following way: Because of complete reducibility, $\mathfrak{P}$ is a direct two-
sided summand, the second summand hence ring-isomorphic to
$\mathfrak{S}/\mathfrak{P}$; and furthermore determined uniquely; therefore equal to
$\mathfrak{S}e_1$, since all conditions are satisfied here.

4. When one passes from $\mathfrak{S}$ to $\mathfrak{o}$, it turns out that both compo-
nents, $\mathfrak{S}e_1$ and $\mathfrak{S}e_2$, are extensions of ideals $\mathfrak{n}$ and $\mathfrak{p}$ from $\mathfrak{o}$;
for both of them have ideal bases from $\mathfrak{o}$ (for instance $N = he_1$
and $M = he_2$, which can take the place of e_1 and e_2 in $\mathfrak{S}$). It
follows that the original ideals $\mathfrak{n}$ and $\mathfrak{p}$ in $\mathfrak{o}$ are contractions of
$\mathfrak{S}e_1$ and $\mathfrak{S}e_2$, respectively, i.e. are the intersection of $\mathfrak{S}e_1$ (respectively $\mathfrak{S}e_2$) with $\mathfrak{o}$ (by more precise consideration of bases,
namely $((a_2 - 1), \ldots, (a_h - 1))$ in $\mathfrak{p}$ and N in $\mathfrak{n}$). Therefore in $\mathfrak{o}$:
$\mathfrak{n} = N\mathfrak{o}$, $\mathfrak{p} = M\mathfrak{o}$. Here $\mathfrak{p}$ also becomes equal to the difference
ideal $((a_1 - 1), \ldots, (a_h - 1))$ in $\mathfrak{o}$. The difference ideal $\mathfrak{p}$ in $\mathfrak{o}$
therefore becomes principal, with the basis $M = he_2$; likewise,
the "difference quotient" (compare 5.) $\mathfrak{n}$ becomes principal
with the basis $N = he_1$ (here h is the corresponding different
$(N)_{a_i \to 1}$). You see that these are the first thoughts from my
lecture in Prague,[102] merely given non-commutatively.[103]

5. If one seeks to identify the set of elements in $\mathfrak{S}$ which are
annihilated by e_1, one obtains $\mathfrak{P} = \mathfrak{S}e_2$; because of $s = se_1 +$
se_2 for every s from $\mathfrak{S}$ and because of orthogonality: $e_1e_2 =$
0; e_i in the center. Similarly $\mathfrak{S}e_1$ consists of precisely those
elements annihilated by e_2 ($\mathfrak{S}e_1 = \mathfrak{S} : \mathfrak{P}$).

On the other hand, in $\mathfrak{o}$ one can only deduce: if an element
t is annihilated by N, then one obtains: ht in $\mathfrak{p}$; and all elements from $\mathfrak{p}$ are annihilated by N because $MN = 0$ and
$ht = Mt + Nt$. Note here the following: the last deductions,
working entirely in $\mathfrak{o}$, remain valid when one passes in $\mathfrak{o}$ from
the integral coefficients to the residue classes with respect to
any integer, as it is indeed the case in the applications. (If
one does not want to do that, then the annihilation by N (respectively M) implies the annihilation by e_1 (respectively e_2);
therefore in 4. the term difference quotient). You work with an
analogue to $h = M + N$ on p. 271, 7.) in Ia);[104] for the cyclic

group $\mathfrak{p}$ also has the basis $(\sigma - 1)$: the group ring $\mathfrak{p}$ becomes the residue ring with respect to $\sigma^l - 1$; respectively $\sigma^{l^n} - 1$.

The application to class field theory I imagine thus:

6. Let K be a Galois extension of k, and $\mathfrak{S}$, $\mathfrak{o}$ the group rings associated to the Galois group of K/k. Let furthermore $\mathfrak{J}$ be any ray class group of K with respect to an ideal modulus $\mathfrak{m}$ from k (or more generally with respect to an invariant modulus in K); hence the hypotheses of Ia, §19, but K does not have to be abelian. $\mathfrak{J}$ has $\mathfrak{o}$ as a multiplier domain, if one writes $\mathfrak{J}$ additively for the sake of comfort (otherwise as exponent domain); here the interpretations are the usual ones: If $A_1, \ldots, A_t$ (as I see, I have unfortunately already used h) are the elements of $\mathfrak{J}$, i.e. the ray ideal classes in K, then $a_i A$ means the application of the substitution a_i to A; and $(a_i + a_j)A$ means the sum (or more precisely the product) $a_i A + a_j A$. The ray A_1 itself hence becomes zero in the additive way of writing. In $\mathfrak{J}$ there is furthermore the subgroup $\mathfrak{R}$ of *rational* ray classes, i.e. those containing ideals from k; $\mathfrak{R}$ contains the ray (zero).

Now the norm of all classes A, in short the norm of $\mathfrak{J}$, can be defined as multiplication (power) with $N = a_1 + a_2 + \cdots + a_h$, i.e. the norm of $\mathfrak{J}$ means the multiplication (power) with the ideal of the integral group ring $\mathfrak{o}$ corresponding to the identity representation. Because of $t\mathfrak{J} = 0$ (the t-th power of every class is the unit of $\mathfrak{J}$) the absolute multiplier ring (§1) is the ring $\mathfrak{o}$ with coefficients modulo t or a residue ring thereof.

7. Since 5. implies: $MN = 0$, where M basis of the difference ideal $\mathfrak{p}$ in $\mathfrak{o}$, it follows immediately: the norm of $\mathfrak{p}\mathfrak{J}$ — i.e. of $\mathfrak{J}$ to the power of the difference ideal — vanishes, lies in the ray. Conversely, however, one can only deduce: if $NA = 0$, then hA is in $\mathfrak{p}\mathfrak{J}$; because of $hA = MA + NA$; $NA = 0$; MA in $\mathfrak{p}A$ in $\mathfrak{p}\mathfrak{J}$, i.e. the h-th power of every class of the principal ray is in $\mathfrak{J}$ to the power of the difference ideal (one can therefore restrict to the direct factor of $\mathfrak{J}$ whose orders are not prime to h). That one cannot deduce more in general already follows from the theorems about the unit principal genus of units (Ia, §12, Satz 12).[105]

The question now arises: is there *in general* a *principal genus theorem* in such a way that for any suitably chosen $\mathfrak{m}$ there is an invariant divisor $\mathfrak{M}$ such that in the class group $\mathfrak{C}$ arising from $\mathfrak{J}$ by complexification to $\mathfrak{M}^{106}$ it holds that: $NA = 0$ (in $\mathfrak{C}$) implies that A in $\mathfrak{pC}$? And here, where the formal stops and the arithmetic begins, I do not know anything. It seems to me, however, that precisely here your new investigations apply, which partly hold for arbitrary K, to understand a bit more than up to now, first in the abelian case and then perhaps generally. After all, in the formal things above $\mathfrak{J}$ can be any abelian group which admits $\mathfrak{o}$, the relation with ideal classes is not at all present yet.[107]

8. I have a few more remarks about possible further approaches, but for the moment they are still pure fantasy. Since in the abelian case the unit of the class group in k to which K belongs, i.e. the ideal group H, arises by taking norms, it therefore corresponds to the identity representation of $\mathfrak{o}$ (multiplication by N). Now the other representations, which arise by direct sum decomposition of $\mathfrak{P} = \mathfrak{S}e_2$ (respectively $\mathfrak{p}$) after extension of coefficients, precisely give the characters; and the group of characters (composition group of representations) is isomorphic to the original group of K/k; should one not be able to obtain the map to the class group in K also from here (besides Artin's map or other interpretations of it) and bring it into relation with your §8, I?[108] This composition of representations also exists in the non-commutative setting, and leads to a commutative ring (Speiser, 1. edition: §45).[109]

Further, I do not think it impossible that in a more strongly arithmetic construction my Prague different theorems should play a role: following them, I can interpret the different as something akin to the conductor of an order (in the Dedekind sense), in the extended hypercomplex system of the maximal order. Only the sum of the components, where therefore $\mathfrak{d}$ is in the denominator, becomes maximal.

Now do what you want with these fantasies: Beste Grüße, Ihre
Emmy Noether.

Notes

[98]Together with Robert Fricke and Öystein Ore, Emmy Noether edited
Dedekind's Collected Mathematical Works [Ded32].

[99]Emmy Noether refers to her return from the DMV-meeting in Prague,
16–23 September 1929. At this meeting Noether had given a talk with the
title *"Idealdifferentiationen und Differente"* (Ideal differentiations and differents). She was however unable to complete the planned detailed publication.
The part which was already finished appeared posthumously in [Noe50]. —
Hasse had given a general talk at the meeting in Prague, which appeared
under the title "The modern algebraic method" [Has30b]. (See also [Has86].)

[100]Apparently Hasse had asked on his postcard what Noether knew about
the connection between hypercomplex algebra (i.e. the theory of algebras)
and class field theory. As one can see from this letter and as Noether herself
remarks, this was indeed still very modest.

The question arises why Hasse asked Noether this question at this point.
According to the date of the postcard, Hasse had written it shortly after the
DMV meeting in Prague where he had met Noether. It therefore seems likely
that Noether, in a conversation with Hasse, had mentioned some ideas about
the theory of algebras in connection with class field theory, and now Hasse
had asked for further details. In this context one may consider the following:

Firstly, Hasse had submitted his paper [Has30c] to *Crelle's Journal* in
March 1929 (seven months before this letter), where he interprets his theory
of norm residues as "local class field theory". This paper marks the birth of
local class field theory. It seems very likely that Noether knew about Hasse's
manuscript, even though the paper had not yet appeared at the time of the
present letter. Maybe Hasse had talked to Noether about it in Prague.

Secondly, Hasse had organized a seminar on the theory of algebras in the
summer term of 1929 (when Noether still was in Moscow). We know about
this from an (undated) letter from Hasse to Kurt Hensel from that term, in
which he writes the following:

> *The scientific center of this term is a seminar on the theory of*
> *hypercomplex numbers, following Dickson's book "Algebras and*
> *their number theory", organized jointly by Jung, Baer and my-*

self. We expect to gain a lot from a thorough understanding of this beautiful new theory, which will certainly be of vital importance for the further development of arithmetic.

Taken together with the fact that Hasse was working on (local and global) class field theory at the time, we infer from this that he was already interested in the connection between the theory of algebras and class field theory in summer 1929. Surely he had also discussed this in Prague with Noether.

Thirdly, Noether had pointed out a possible connection between local class field theory and the theory of algebras in the winter term 1929/30 (that is the term following this letter). This is mentioned at the end of the lecture notes [Noe83a], which were worked out by Deuring at the time (albeit published only posthumously). We also refer to the later letter from 22 November 1931, in which Noether mentions her lecture notes 1929/30 in connection with local class field theory.

Fourthly, Noether cites Hasse's paper [Has31d] in the lecture notes mentioned above, even though this paper was published in 1931 and had been submitted to the *Mathematische Annalen* in June 1930. Noether therefore must have known the contents of this paper before it was finished. In this paper Hasse determines the types of local skew fields. Compare also the letter from 25 June 1930, which directly refers to Hasse's submitted paper and at the same time to the relation with class field theory.

In this light it indeed seems plausible that Hasse and Noether talked about the topic of algebras and local class field theory in Prague, and that the present letter can be interpreted as a first sign of the development of ideas which would later culminate in Hasse's paper [Has33b], when Artin's reciprocity law in class field theory could be proved with algebra-theoretic tools.

[101]Here Noether cites her own major paper "Hypercomplex quantities and their representation theory" [Noe29], which had just appeared.

[102]Here Noether refers to her talk on differents at the DMV annual meeting in Prague in September 1929.

[103]Here Noether adds a footnote: *"Hopefully you can read 4.; it does not work with simplified hints to the proof! It all becomes wrong!"* With this she apologizes for the somewhat unreadable presentation in this section, in which there is a lot of crossed-out text and almost illegible insertions. Apparently she had first only wanted to give a simplifying proof sketch and later decided to be a bit more precise.

[104]Noether means section (7.), page 72 in Hasse's class field report part Ia [Has27a]. The page number 271 should hence be replaced by 72. In this

section (7.), Hasse treats Hilbert's ramification theory, particularly the action of the decomposition and inertia group on the prime ideals of the Galois field extension.

[105]This refers to §12 of part Ia of Hasse's class field report Ia. In Satz 12, which Noether cites, the order of the first cohomology group of the units of a number field is calculated, and it turns out not to vanish.

[106]Noether uses the terminology "complexification" for the group of residue classes $\mathfrak{J}/\mathfrak{M}$

[107]For an arbitrary $\mathfrak{S}$-module $\mathfrak{J}$, Noether's thoughts mean precisely that the first cohomology group $H^1(\mathfrak{S},\mathfrak{J})$ is annihilated by the group order. Here we see the first approaches to an algebraic cohomology theory. In the particular situation described by Noether she asks whether there is a (natural) factor module of $\mathfrak{J}$ in which the first cohomology vanishes. — The term "principal genus theorem" (Hauptgeschlechtssatz) is used by Noether in a rather general sense; it concerns the vanishing of first cohomology groups or certain subgroups thereof. The theorem is closely related to the local-global principle for algebras, which (in today's terminology) is equivalent to the vanishing of the first cohomology of the idele class group. Noether later explained her general ideas in this connection in her Zürich lecture [Noe32a], published in detail in the *Mathematische Annalen* [Noe33b]. In the present letter we see the beginnings of these considerations.

[108]"Artin's map" means assigning to the unramified prime ideals $\mathfrak{p}$ of the ground field their Frobenius automorphisms in the Galois group of the extension. In §8 of part I of Hasse's class field report [Has26a], the theorem of the arithmetic progression is treated under the aspect of class field theory.

[109]The "non-commutative" in this context means that the Galois group does not need to be abelian; this goes beyond class field theory, which only refers to abelian field extensions. — "Speiser" refers to his textbook [Spe27] about group theory.

3.18   Letter of October 07, 1929

Hasse goes to Marburg. Inverse theorem of class field theory.

Lieber Herr Hasse!

First of all my warmest congratulations for your appointment to Marburg! I knew of the nomination, but not of the appointment.

It seems certain to me that you will accept,[110] and hence I already have a few wishes.

Firstly, if you should have been granted a post for an assistant, I would be very glad if you would consider Levitzky. At present he has traveled home, for the first time in seven years; but it is very possible that he should return at some point in winter, and he will surely come if he has the prospect of a post.[111]

Further there is the question whether it is possible to get Aleksandroff on the list in Halle. I know that Aleksandroff strongly wishes to come to a German university some time. And since all his papers are published in German journals – or in German language in America –, perhaps the foreigner will not create any major difficulties, especially since his scientific importance is undisputed. Furthermore he is now writing his "Topology" for the yellow collection; and finally he has already received payment from the government for the Göttingen guest lectures for two summers, even with tax deduction. Incidentally, besides topology he is lecturing Galois theory this winter, of course in the modern way; and he is always intensively working in seminars with his people. Also papers of various students of his have appeared in the Annalen, and further ones are going to come. You will know that he has a complete and flawless command of the German language.[112]

That would be it for my wishes for foreigners!

Your Kalkutta-Himalaya comparison is very right. If it will be possible to use hypercomplex approaches to get into the arithmetic parts of the inverse theorem,[113] then I believe that these approaches will be in the further development of my Prague different theorems[114] — in connection with the formal theory of the group ring. But when will one be this far? The possibility itself is not at all proven — and yet the inverse theorem as a structure theorem seems to have nothing to do with analytic methods! [115] In any case I hope to be able to learn some things from your recent papers.

Beste Grüße, Ihre Emmy Noether.

Notes

[110]Hasse had been appointed to Marburg as the successor to Hensel. Hensel had been Hasse's "Doktorvater" (PhD supervisor), and was now his "paternal friend" (according to Hasse's own words). Hence Noether (and all other mathematicians who knew Hasse) could well assume that Hasse would accept the appointment to Marburg. He started in the position at Easter 1930.

[111]Regarding Levitzky see the postcard from August 12, 1929. 3.16

[112]The Russian mathematician Paul Aleksandroff (1896–1982) was frequently Emmy Noether's guest in Göttingen. And Emmy had been in Moscow in 1928/29, at Aleksandroff's invitation. His works on algebraic topology were strongly influenced by Noether, especially the ground-breaking book "Topology" also mentioned by Noether, which he published together with Heinz Hopf [AH35]. That he would have liked to take a chair in Germany at the time was not generally known, and is only evidenced by the present letter of Emmy Noether. Incidentally, Noether's suggestion was not followed up. Hasse's successor in Halle was Heinrich Brandt, who had distinguished himself by his works on the arithmetic of algebras. — Some letters from Noether to Aleksandroff have recently been found; they have been edited by Renate Tobies. [Tob03]

[113]This concerns the inverse theorem of class field theory. This is due to Takagi and states that every relatively abelian field can be described as a class field.

[114]Here Noether refers once more to her talk at the DMV meeting in Prague in 1929, published in [Noe29].

[115]A direct method to prove the inverse theorem without using analytic tools was given much later by Chevalley and Nehrkorn [CN35].

3.19   Postcard of November 13, 1929

Deuring's dissertation. Galois theory of skew fields.

Lieber Herr Hasse!

I wanted to ask whether you will soon receive corrections for your new class field theory papers?[116] One of the best students here, who has on his own initiative studied class fields in the case of function fields in one variable with rational algebraic coefficients (i.e. characteristic *zero*) is eagerly waiting for the delivery promised to me. – He considers subgroups of finite index in the *infinite* group of divisor classes, and the corresponding factor groups, the class groups. In the case of algebraically closed coefficient domain the results are almost trivial; everything is realizable by pure equations.[117] But he also has approaches for rational coefficients: transfer of the Riemann–Roch Theorem and analogue to quadratic reciprocity. The Riemann–Roch Theorem will presumably have to replace transcendental methods as far as possible.

At the same time I wanted to ask you whether there were any prospects for Levitzky in Marburg; otherwise I could potentially try to do something for him at the inaugural celebration of the institute, where a lot of people will come. Are you also going to come? (3 December) The invitation cards are presumably still due to be sent.[118]

After remarks of Van der Waerden, I can now greatly simplify The Galois theory of non-commutative fields, and extend to two-sided simple hypercomplex systems. I am probably going to publish it in this way.[119]

Beste Grüße, Ihre Emmy Noether.

Notes

[116]It is not clear which papers Noether had in mind. In 1930, six papers of Hasse appeared which concerned class field theory.

[117]The student whom Noether writes about is Max Deuring. He earned his doctorate under Noether in 1930 with work on class field theory of algebraic function fields in one variable [Deu31b]. However, this does not concern subgroups of finite index, but finite subgroups of the divisor class group. – For function fields with algebraically closed constant field, Kawada and Tate [KT55] later described Deuring's theory within the framework of cohomology theory, at least in the unramified case. For the tamely ramified case see Madan [Mad66].

[118]On 3 December 1929 the new building of the Göttingen mathematical institute, financed by the Rockefeller foundation, was inaugurated. It is unknown whether Hasse was present at the celebration.

[119]See [Noe33b].

3.20   Postcard of June 25, 1930

Hasse's hypercomplex p-adic. Local class field theory.

Frankfurt/M.[120]

Eschersheim

Haeberlinstr. 53

Lieber Herr Hasse!

Your hypercomplex p-adics have pleased me immensely.[121] I will be able to hand them on to Blumenthal[122] on approximately 7 July, since he will be back only then from the Kharkov congress;[123] but the *Annalen* print very quickly at the moment. Would you not want, in the proof stage, to insert a remark at the last §about the similar deductions in Prüfer and v. Neumann (your name "component" for $\mathfrak{a}_p$ already hints at this).[124] This method now also repeatedly appears in Köthe,[125] Pietrkowski[126] and so on. One should mention this explicitly.[127] — By the way, I do *not* believe that Prüfer-v. Neumann suffice for the passage from local class field theory to global class field theory; as you do not get theorems about ideal *classes* even now.[128] I believe that one will here need to work in parallel with the ideal group instead of with field elements, as Dedekind (Gött. Nachr. 93) has already done in the theory of modules.[129]

In this context: it follows from local class field theory: if Z is cyclic of degree n over a $\mathfrak{p}$-adic base field K, then in K there is at least one element $a \neq 0$ such that only a^n becomes the norm of an element of Z. Can you prove this directly? Then one could conversely derive local class field theory from your results on skew fields; but I have only thought about the first line of the Isomorphism Theorem 1, p. 147 in your paper; the rest is just a claim.[130]

Beste Grüße, Ihre Emmy Noether.

Notes

[120]In the summer term of 1930 Emmy Noether stayed in Frankfurt am Main as a visiting professor of the University there — in exchange with Siegel who lectured in Göttingen for this term. I do not know the subject of Noether's lecture in Frankfurt. Siegel's lecture in Göttingen on analytic number theory was worked out by Noether's student Deuring. Emmy Noether wrote from Frankfurt in a postcard to Deuring: *"It is very nice that you are doing the notes for Siegel; then I can calmly read his breakneck proofs, which I prefer to listening."* (This is reported by Martin Kneser in his obituary of Deuring [Kne87].)

[121]Hasse had sent his paper "On ℘-adic skew fields and their significance for the arithmetic of hypercomplex number systems" [Has31d] for publication in the *Mathematische Annalen*. Emmy Noether worked as an (unofficial) editor of the *Mathematische Annalen*. The paper carries the submission date 18 June 1930. Therefore Noether read the manuscript within one week. Probably Hasse had told her earlier about the content of this paper. Compare the remarks to the letter from 2 October 1929.

[122]Blumenthal was editor-in-chief of the *Mathematische Annalen*.

[123]The University of Kharkov had been founded in 1805. The "Mathematical Society of Kharkov" was founded in 1879.

[124]Prüfer's paper "A new foundation of algebraic number theory" [Prü25] had appeared in the *Mathematische Annalen*. The reference to v. Neumann is to the paper "On Prüfer's theory of ideal numbers" in the Acta Szeged [vN26]. Hasse followed Noether's suggestion, as in §8 of his paper both Prüfer and v. Neumann are cited.

[125]Noether probably refers to the paper "Abstract theory of non-commutative rings" [Köt30].

[126]"Theory of infinite abelian groups" [Pie31a]. "Subgroups and quotient groups of infinite abelian groups" [Pie31b]. These papers had appeared in the *Mathematische Annalen*.

[127]The method of proof used in § 8 of Hasse's paper which Noether mentions consists of the passage from local to global, or more precisely the characterization of global domains by their local components. For instance it is shown that a (right or left) ideal of a maximal order is uniquely determined by its local components; and also that there always exists a global ideal corresponding to a given system of local ideals satisfying certain natural conditions, and so forth. Noether sees all this as part of a general method, which would be useful in other situations too.

[128]Hasse had in mind that one should find methods for the passage from local to global class groups. At that time one had not been able to do this in a satisfactory manner.

[129]Dedekind had published two papers in the Göttinger Nachrichten in 1895. It appears that Noether means the second one [Ded95], with the title "On an extension of the symbol $(\mathfrak{a}, \mathfrak{b})$ in the theory of modules". There Dedekind gives a module-theoretic foundation for the theory of relative norms of ideals. In the collected mathematical works [Ded31] of Dedekind, Noether remarks that it *"has not been applied to global arithmetic questions"*, and adds that the paper might yet become important for the theory of field extensions.

[130]This refers to the isomorphism theorem $\overline{1}$ in Hasse's paper "The theory of norm residues as local class field theory" [Has30c] in *Crelle's Journal*. The first line of the theorem in question says (with the notations used by Noether): *The Galois group of Z/K is isomorphic to the norm class group $K^\times/NZ^\times$.* This theorem had been proven in the context of local class field theory by Hasse [Has30c] and F. K. Schmidt [Sch30] (published in *Crelle's Journal*), but the proof given there relied on *global* class field theory. More precisely, the isomorphism is given by the $\mathfrak{p}$-adic norm residue symbol, which Hasse could not define locally at the time, which is why he had to use global class field theory. Here Noether asks for a "direct" proof, thus without recourse to global class field theory.

The proof can indeed be given directly, as an immediate consequence of Hasse's theory of $\wp$-adic skew fields, which Noether refers to in the present letter. But it appears that Hasse himself did not at once see this way and therefore answered Emmy Noether's question with "no" or "I do not know"; we deduce this from a remark by Noether in her later letter from 2 May 1932.

Noether's question comes down to whether one can choose $a \in Z$ in such a way that the crossed product (a, Z, S) has the precise exponent n; here S denotes a generator of the Galois group of Z/K, where the choice is inconsequential. Now Hasse proves in his paper [Has31d] on $\wp$-adic skew fields that there exists a distinguished division algebra over K of exponent n, namely the crossed product $D = (\pi, W, F)$, where π is a prime element of K, W the unramified extension of K of degree n and F the Frobenius automorphism of W/K. Accordingly Noether's question comes down to whether Z is a splitting field of D, for then Z can be embedded into D, and therefore D can also be written as a crossed product with Z: $D = (a, Z, S)$ with suitable $a \in K$, which has exact order n with respect to the norm factor group from Z.

Later, namely in the joint paper [BHN32] about the main theorem of the theory of algebras, it is shown that *every* extension field of K of degree divisible by n is a splitting field of D (Satz 3); this proof is due to Hasse alone, and for this he uses the results of the paper discussed here [Has31d]. Hence Hasse could have answered Emmy Noether's question positively at this point already.

However, the foundation of full local class field theory building on this turned out to involve some difficulties in the passage from cyclic to abelian extensions. Noether admitted this in her talk in Zürich in 1932, when she says there that *"new algebraic theorems about factor systems still had to be developed"*. Chevalley did this in a paper in *Crelle's Journal* [Che33a]. From today's perspective this concerned basics of cohomology theory, in particular restriction, inflation and transfer. Compare here the letter from 2/3 June 1932.

3.21 Letter of October 10, 1930

Artin's conductors and theory of algebras.

Lieber Herr Hasse!

Many thanks for Artin![131] The things are really very nice! I am particularly pleased about the formal foundations in it; I have thought of some hypercomplex things — entirely independent for now — such as the following:

The crossed product of K with its group ring (group of K/k) becomes a full matrix ring because of factor system *one*. Every basis of K/k — together with the unit of the identity representation of the group ring — yields a decomposition in one-sided simple, say right ideals. The corresponding left decomposition is then generated by the complementary basis of K/k. If one restricts to integral ideals, then right and left decomposition hence belong to complementary ideal classes of K/k. I suspect that one also obtains theorems about decomposition of the different in this way, and then after taking norms relations with Artin. But those are dreams for the future! [132] In any case many thanks for showing me the letter; sending it back was delayed a bit as Deuring was away travelling, to whom I wanted to show the letter. Incidentally, he has recognized something very similar to the above as the formal reason for his exchange theorem: passage to the reciprocally isomorphic ring (exchanging right and left multiplication) together with exchanging field and group in the cyclic case, when one puts crossed product with arbitrary factor system as the base. [133]

Beste Grüße, Ihre Emmy Noether.

Notes

[131] Hasse had sent Noether a letter which he had received from Artin. That letter was dated 18 September 1930 and contained Artin's addition to his theory of L-series, in particular the theory of Artin conductors. Compare [Art30].

[132] A part of these dreams were later realized by Noether in her paper about integral normal bases in the tamely ramified case, which appeared in the 1932 volume of *Crelle's Journal* dedicated to Hensel [Noe32b].

[133] Here Noether refers to Deuring's algebra-theoretic proof of the exchange theorem for the Hilbert norm residue symbol [Deu31c].

3.22   Postcard of November 02, 1930

Legacy of Fricke. Different. Criterion for unique factorization.

Lieber Herr Hasse!

Indeed it appears that Teubner has simply not yet written to Koschmieder. Fricke's daughter — Frau Landauer, Braunschweig, Kaiser-Wilhelmstr. 17 — asked me for advice regarding Teubner's suggestion "Koschmieder", and when I agreed she wrote to me that she and her brother now wanted to finalize everything with

Teubner. This was about three weeks ago; perhaps it is easiest if you now directly approach Frau Landauer to hear how the matter stands.[134]

The ideal different is promised to you.[135]

Courant told me about Jena, and also Krull wrote to me from there with satisfaction. He apparently told you that he has now transformed your criterion for decomposition into prime elements into a necessary and sufficient one, by simply adding a factor prime to a![136] Incidentally, can one not shorten Wegner? After all, he has three papers about the subject in total.[137]

Deuring would like to know whether he will get a copy of the report, since he would want to buy it otherwise; he would prefer a free copy![138] (Hospitalstr. 3a). I do not think it impossible that with his new approaches one can also get to the question of the 5 axiom rings: this is about the construction of general valuations.[139]

Herzliche Grüße, Ihre Emmy Noether.

Notes

[134]Robert Fricke held a chair at Braunschweig University, as the successor of Richard Dedekind. Jointly with Emmy Noether and Öystein Ore he edited the collected works of Dedekind [Ded32]. Fricke died in 1930 during the preparation of the third volume. Apparently Noether and Hasse advised Fricke's daughter about the handling of his mathematical estate. Since the name of the mathematician Koschmieder had been mentioned by the publishing house Teubner, it appears that he was discussed as an editor of yet unpublished parts of Fricke's estate.

[135]At the annual meeting of the DMV in September 1929 in Prague, Noether had given a talk about "Ideal differentiation and different". Apparently Hasse had asked for a preprint. But Noether did not finalize the corresponding manuscript during her lifetime. Before she had to leave Germany she gave the manuscript to her former student Heinrich Grell. It appeared after the war in the year 1950 in *Crelle's Journal* [Noe50].

[136]For Hasse's criterion compare the letter of 11 December 1929. Krull's criterion can be found in [Kru31] on page 11, Satz 2. Krull says: *"A simple*

modification of Hasse's criterion leads to a necessary and sufficient criterion for the decomposition of prime elements". Krull's criterion reads: *An integral domain R is a unique factorization domain if and only if there exists a non-trivial additive function* $\chi\colon R \to \mathbb{N}$ *with* $\chi(ab)) = \chi(a) + \chi(b)$ *which satisfies the following condition: if* $a, b \in R$ *with* $a \not\equiv 0 \mod b$ *and* $b \not\equiv 0 \mod a$ *then there exist* $r, s, q \in R$ *with* q *prime to* a *and*

$$ra + sb = qc \quad \text{with} \quad \chi(c) < \min(\chi(a), \chi(b)).$$

(Krull uses the additive form for χ as Noether had already proposed in her letter of 11 December 1929.) The essential generalization of Hasse's criterion is the factor q in the above condition.

[137]This refers to Udo Wegner. In volume 105 (1931) of the *Mathematische Annalen*, Wegner had seven papers published. In the following volume 106 a note by Hasse appeared, in which he showed that the main result of one of these papers is not correct [Has32e]. (This concerns the reducibility of polynomials which decompose almost everywhere.) Hasse there refers to an oral communication from Artin in 1927, who had produced a counter example to the theorem of Wegner. Hasse then gives a correction of Wegner's theorem.

[138]This is the second part of Hasse's class field report which had just appeared [Has30a].

[139]Noether's name for Dedekind rings is "5 axiom rings". Noether's five axioms are found in her paper [Noe24]. Apparently Deuring was working on characterizing Dedekind rings and their ramification valuation-theoretically. Compare [Deu31a].

3.23   Letter of December 19, 1930

Hasse's conjectures on algebras. Albert's alleged counterexamples.

Göttingen, 19 December 1930

Lieber Herr Hasse!

Yes, it is very sad that all your nice conjectures are only floating in the air and not standing on the ground with firm feet: for a part of them — I do not quite see how much – falls hopelessly to the ground because of counterexamples in a new American paper: Transactions of the Amer. Math. Society, vol. 32; by Albert.

This implies firstly that the exponent can really be *smaller* than the index, already with the rational number field as center; and therefore further that your theory of forms *cannot* be transferred to forms of higher degree. Whether your conjecture about the cyclic splitting fields holds is at the very least dubious.[140]

More precisely this is about the following: Albert proved in vol. 31 [141] that all skew fields of index 4 with rational center — if any such exist — are direct products of two skew fields of index 2; which immediately implies that they have exponent two, in exact analogy with Brauer's example.[142] In vol. 32 [143] he now gives, page 188, explicitly an example of such a skew field; hence $n = 4$, $q = 2$. Further, into this a cyclic field of degree 4 can be embedded in such a way that already the square of β becomes a norm; Dickson[144] is sufficient, but not necessary.[145]

Further he gives necessary and sufficient conditions in the case of $n = 4$ for there to be *no* cyclic embedded field of degree 4; but it is not said whether there are solutions here over the rational numbers. But very possibly you can easily understand the first, algebraic, i.e. form-theoretic form of the conditions, and hence construct a counterexample, whose existence does not seem at all impossible to me.

Further numerical material can be found in the paper by Archibald cited in the introduction, at the very end. Archibald is here this semester; and this is also the origin of my knowledge of American literature. But he does not know whether there are any systems without zero-divisors among his collection.[146]

Do let me know if also your cyclic splitting field should come to nothing!

I now want to form a crossed product of the (absolute) class field with its group; then in the solvable field generated by the group I obtain, with a suitable choice of the factor system — the $s_i = \sigma_i^{k_i}$ corresponding to the cyclic factors of the group —, Hecke's ideal numbers; the s_i need only be put equal to the k_i-th power of the associated ideal class. In the ambiguity of the s_i, caused by the units, and the corresponding ambiguity of the skew field, units, principal genus and so forth must play a role! But this is all nebulous for now! [147]

Beste Grüße, Ihre Emmy Noether.

Notes

[140] Apparently Hasse had informed Noether about a series of systematically founded conjectures in the theory of algebras over number fields. We do not know the precise statements of these. But we know from Noether's letter that at least the following three conjectures about algebras over number fields were among them:

1. The index of a simple algebra is equal to its exponent.

2. The local-global principle for the splitting of simple algebras. (Noether uses the term "theory of forms", following Hasse's theory of quadratic forms, for which he had proven a local-global principle in his dissertation; here the norm forms of simple algebras are meant.)

3. Every simple algebra has a cyclic splitting field.

Hence this letter by Noether is an important historical document. We can infer from it approximately when Hasse explicitly stated these conjectures, namely in December 1930. It was an error when Noether called a part of Hasse's conjectures invalidated by a counterexample; Hasse immediately pointed this out to her. In the next letter of 24 December she herself admits it. — By the way, it was Emmy Noether herself who had given Hasse a hint that there may be a local-global principle for the splitting of algebras: During one of Hasse's frequent visits in Göttingen, both went for a walk to the hill Hanstein in the vicinity, and there Hasse told her that the local-global principle for norms holds for cyclic fields, but not in the general abelian case. Noether reacted by proposing that instead of norms one should consider splitting factor systems. It appears that this provided the stimulus for Hasse to conjecture the local-global principle for algebras. Compare Noether's letter of 12 November 1931, where Noether refers to "Hanstein".

[141] This refers to volume 31 of the Transactions of the American Mathematical Society (1929) with the paper [Alb30]. (Attention: In volume 31 of the Annals of Mathematics (1930) there is also another paper by Albert with a similar subject, namely quaternion algebras over quadratic number

fields [Alb31b]. However, Noether appears to refer to the first paper, which contains essentially Albert's dissertation.)

[142]We do not know Brauer's example which Noether has in mind. But she is mistaken when she claims that it is a counterexample to Hasse's conjecture. For, Brauer's examples as presented in [Bra33] are all given as division algebras over function fields of higher transcendency. Although that paper of Brauer's had not yet appeared at the time of this letter, it may be that Brauer had informed Noether about its content or at least part of it. But Noether did not realize that the base fields of Brauer's examples are all of higher transcendency.

[143][Alb30].

[144]Here, "Dickson" means the norm criterion for a cyclic algebra being a division algebra. Let a crossed product $A = (Z, \sigma, \beta)$ of a cyclic field Z/K of degree n with a generating automorphism σ and a parameter $\beta \neq 0$ in the ground field K be given. Then the norm criterion says: If n is the smallest exponent for which β^n is a norm from Z, then A is a division algebra. Compare Dickson, "Algebras and their number theory" [Dic23], §42. Hasse in his American paper "Theory of cyclic algebras over an algebraic number field" [Has32c] speaks of "Dickson's criterion". However, the criterion is in fact due to Wedderburn [Wed14]. If the German mathematicians refer to this as "Dickson's criterion" then this is a sign that Wedderburn's papers were not well known in Germany and that Wedderburn's results on the structure of algebras were disseminated mainly through Dickson's book.

[145]Later, Emmy Noether noted herself that she was in error here; compare the next letter of 24 December 1930. If the base field of the division algebra is a number field then the norm criterion is indeed not only sufficient, but also necessary.

[146]This concerns [Arc28]. This paper is cited in the introduction of Albert's paper referred to by Noether. — Archibald stayed in Göttingen at the time and was in mathematical contact with Emmy Noether.

[147]At the time of this letter, the principal ideal theorem of class field theory had already been proven: [Fur29]. Noether therefore knew that every ideal of a number field becomes a principal ideal in its absolute class field. The generators of these principal ideals are apparently what Noether calls "Hecke's ideal numbers". Noether here appears to attempt to explicitly construct these ideal numbers with the aid of a split crossed product. It is not clear whether she was aiming for a new, "hypercomplex" proof of the principal ideal theorem. Noether's idea has apparently not been developed further.

3.24   Letter of December 24, 1930

The boat picture. Correction to last letter.

Göttingen, 24 December 1930

Lieber Herr Hasse!

Attached you find the metamorphosis of your boat picture[148] into a negro woman — from the Mapha film;[149] but apart from that, this letter is a pater peccavi. For your castles in the air have not yet collapsed — perhaps you are now sad about that — I have read in Albert's "monstrosity"[150] quite the opposite of what it said. Only your counterexample clarified the matter for me.

Hence: (1) In the first paper (Transactions 31) it is shown that every skew field with rational center of index 4 (abbreviated skew field) has a maximal abelian subfield of degree 4.

(2) In the second paper (Transactions 32) the consequent representation as a crossed product is discussed; and the diophantine conditions for cyclic generation are given.

(3) In a third paper (Ann. Math. 30 (1929), p. 621) it is shown that the direct product[151] of two generalized quaternion fields always contains zero divisors; they are constructed by means of the theory of quadratic forms.

My mistake was to deduce direct product from (1) and (2). In fact the matter is as follows:

Let $Z = R(\alpha_1) \otimes R(\alpha_2)$ be the Abelian subfield; $G = \{1, \beta_1, \beta_2, \beta_1\beta_2\}$ the corresponding group; with $\beta_1^{-1}\alpha_1\beta_1 = \overline{\alpha}_1$ and $\beta_2^{-1}\alpha_1 \beta_2 = \overline{\alpha}_2$; $\beta_1^2 = b_1$, $\beta_2^2 = b_2$, $\beta_1\beta_2 = c\beta_2\beta_1$. Then b_1 and b_2 do not have to be rational[152] at all, and c does not need to be associated to the unit (i.e., become 1 by passage to generators β_i'); this I had overlooked, as I did not write down the "factor systems" b_1, b_2, c.[153]

One can only deduce: b_1 admits the transformation with β_1, and hence is in $R(\alpha_2)$; likewise b_2 in $R(\alpha_1)$. (Between b_1, b_2, c and conjugates there further is an associativity condition.) In fact b_1 and b_2 must now be *nonrational*: this is shown in 3). (Likewise $(\beta_1\beta_2)^2$, which is expressible by b_1, b_2, c, of course not rational; as a subfield with same center always direct factor). This really is the opposite of what I recently claimed!

It therefore does seem rather likely that with algebraic number fields as center, exponent and index always agree. Nevertheless using the Abelian splitting fields seems right to me in principle. I suspect that for non-zero divisors the conditions of the character of "not rational" are more easily satisfied, which complement the condition "non-norm", if the matter is more complicated that in the p-adic situation.[154]

Such extra conditions are of course necessary when — as in Brauer's example — the exponent becomes smaller than the index by taking direct products; the non-norm conditions only say something about the exponent, after all. However, Brauer's construction of his example is based on the fact that the center $R(x, y)$ is isomorphic with the two quadratic extension fields $R(x, y, \sqrt{x})$ and $R(x, y, \sqrt{y})$, that is with $R(\sqrt{x}, y)$ and $R(x, \sqrt{y})$. This cannot occur in the absolutely algebraic situation.[155] Perhaps such an isomorphism is even necessary to push down the exponent, I do not know!

Hence you see that your question about when direct products of skew fields are again skew fields is not an easy one. With coprime degrees this is always satisfied; this was proved by Köthe and Brauer; perhaps this is also known to you.[156]

With best wishes for Christmas and the new year, for humans and skew fields, Ihre Emmy Noether.

Notes

[148]The "boat picture" was a photograph of Emmy Noether, which Hasse had taken in September 1930 during a passage over the Baltic Sea to the annual meeting of the DMV in Königsberg (Kaliningrad). — The picture is

published in Auguste Dick's Noether biography [Dic70]. On 3 May 1930 Mrs Dick sent a copy of the biography to Hasse; in her accompanying letter she wrote:

> ... *The photo on the title page presumably pleases you? I had also offered, to the publisher, a picture from the archive of the university of Göttingen, which was taken by a professional photographer. Emmy Noether in solemn posture. But your picture was preferred and I am glad about it...*

We have reproduced the boat picture in this book. — Regarding this picture compare also the letter of 2 December 1931.

[149] "Mapha" was the name of an association of mathematics and physics students in Göttingen. Apparently on the occasion of a student party, Mapha had created a photo-montage by placing Emmy Noether's head from the "boat picture" on the body of an African market woman. Possibly they had created similar joke pictures for other professors too.

[150] Noether used the German "Fürchterlichkeit". Apparently Noether was indicating her frustration with Albert's exposition. This word apparently indicates that Noether disagreed with Albert's style. In his early papers Albert's presentation was similar to that of his doctoral advisor, Dickson. This was very different from the abstract style of Noether and her students.

[151] In today's terminology, in the present situation one would say "tensor product" instead of "direct product".

[152] Here "rational" means "contained in the base field".

[153] The group G is the direct product of the groups $G_1 = \langle \beta_1 \rangle$ and $G_2 = \langle \beta_2 \rangle$. The second cohomology of G, however, is not the direct product of the second cohomology of G_1 with the one of G_2, but one has to use the Künneth formula. The factor c represents the additional term from the Künneth formula. Note that in Noether's times there was not yet a systematic cohomology theory for groups; this developed step by step from the experiences of calculations with factor systems. Here we see an example for Noether's approach to such experiences.

[154] Hasse's paper about $\wp$-adic skew fields [Has31d] had not yet appeared at the time of this letter, but Noether knew its content. Compare the letter of 25 June 1930. Hasse had shown in the paper that over a $\mathfrak{p}$-adic field every skew field is a cyclic crossed product. — Noether apparently wished to find a criterion that a crossed product with *abelian* Galois group does not have zero divisors, i.e., it is a skew field. Albert had given such a criterion in the form of unsolvability of certain diophantine equations, but this is difficult to verify in individual cases. The final solution of the question — and not

only for abelian crossed products — was eventually solved through the *Hasse invariants* of algebras [Has33b] — at least when the base field is a global field. For, the exponent of an algebra is the least common multiple of all its Hasse invariants; if this is equal to the degree of the algebra, then the algebra is a skew field.

[155]Brauer's example can be found in his paper [Bra30]. There the base field is the rational function field in two variables over $\mathbb{Q}$. — Later Albert [Alb32a] discussed the same construction, and he claimed that Brauer's argument is false in general. However, he admits in a footnote at the end of the paper that this was a problem of "interpretation of language, rather than a mathematical error". In fact both papers are correct, Brauer's as well as Albert's. Albert's constructions are more general than Brauer's. Albert succeeded in the construction of non-cyclic division algebras of index 4, whereas Brauer had answered the related but different question whether there are algebras of index 4 whose exponent is smaller than the index. — Remark: In [Cur99] it is erroneously claimed that Brauer [Bra30] was the first to construct a non-cyclic division algebra of index 4, and this is repeated in [FS05]. In fact this question was not discussed in [Bra30], and priority is due to Albert.

[156]Köthe's paper [Köt31] appeared in the *Mathematische Annalen*. This paper was however only written after Brauer's paper [Bra30], in which the theorem is likewise stated. If Noether now first cites Köthe and not Brauer, then this may be because she was better informed about Köthe's work, as Köthe was in Göttingen at the time. — The same theorem was independently proved by Albert around the same time [Alb31b]. Incidentally: If the base field is a number field, then the converse of the theorem also applies; this follows for instance from the later results in the joint paper of Brauer—Hasse—Noether [BN27].

3.25 Letter of February 08, 1931

Fitting. Skew-Congress in Marburg.

Lieber Herr Hasse!

In response to the questions from your post card you will get much hypercomplex material to work on, perhaps more than you would like.

Fitting has stated his structure theorem on p. 4 as Satz 1; but to understand it, you will first need to work through the auto-

morphisms[157] — for my taste he has presented them excellently
(a field does not need to be commutative). You will also see
how Artin's theorems fall out by specialization (Satz 3 and 4).[158]
Incidentally, in the hypercomplex case his Satz 1 (resp. 2) is simply the interpretation of Satz 10, p. 130 in the German edition
of Dickson[159] but the matrix representation is obtained merely
through the automorphism ring. I was very pleased with Fitting's
work which is *completely independent* in the question and the ex-
ecution — even if influenced by my definitions — and contains
much more than the results stated above; in May he will finish
his doctorate with this, incidentally he already has 11 semesters.
The passage from Satz 1 to Satz 2 corresponds to my passage to
K (p. 667, or Dickson p. 119 (12)).[160]

The answer to the question regarding maximal subfields you can
get out of Van der Waerden (the previous chapters are already
available in proof), who had greatly simplified my proofs from
the summer lectures 1928; I have presented them similarly in
29/30 [161] (but with introduction of the "reciprocal representation
module") instead of p. 117ff (and also much more general apart
from that, as I now notice; but that also partially with Van der
Waerden's simplifications.) You will find the theorem on p. 122,
and therefore only need to read 117–122; I have sent you the whole
manuscript as I do not know whether the notation is comprehensible independently. — As I now see, Van der Waerden has more
than I knew, namely the theorem at the bottom of p. 121.[162] I
think R. Brauer talked about this theorem in Königsberg.[163]

By the way, Van der Waerden also has Fitting's construction of
the Artinian ring (54–56 in the new version, above). This is due
to Rabinowitsch in Moscow, but that was not known to Fitting.

And now regarding the "skew congress".[164] I will give "Hypercomplex structure theorems with number-theoretic applications",
to give a name to the child. By the way, I really hope to be able
to say something about conductor theorems: I have some first approaches, namely for the different decomposition, which is more
easily accessible in a hypercomplex way, from which the Artin

theorems would follow by taking norms. I do not know, however, if I will get that far, otherwise the old applications will do!

I would suggest as the order of talks: R. Brauer, Noether, Deuring, Hasse. Then everyone can refer to the preceding speaker. The other talks are independent.

I also want to sugggest inviting my Rockefeller scholar for the next summer — who is currently with von Neumann in Berlin: Dr. J. Herbrand, Berlin-Charlottenburg, Mommsenstr. 47, c/o Ehrmann. As Rockefeller scholar his travel expenses will be paid; hence there are no expenses for you. He came to Halle, and has understood the most of my things among all.[165] Apart from logic he has only worked on number theory until now (which he has learned from your "report" and your theory of norm residues[166]); I only thought of him as a participant. However he would be able to talk about his integral representations of the Galois group coming from the unit groups; this is probably closely related to my hypercomplex things (C.R. January.[167].) We all had an excellent impression of him in Halle.

Beste Grüße, Ihre Emmy Noether.

Notes

[157]This concerns the dissertation of Noether's student Fitting. It was published in extended form in the *Mathematische Annalen* [Fit32]. The version which Noether refers to here is different from the published version; the page and theorem numbers given here by Noether do not match those in the published version. — The aim of Fitting's paper is a "New foundation of the theory of hypercomplex systems, more generally of arbitrary rings with double chain condition". The new method consists of viewing the ring as a group with itself as a domain of operators; this becomes a special case of the general structure theory of endomorphism rings of groups. (Fitting uses "automorphism" where today one would say "endomorphism".) Apparently Hasse had heard of this dissertation and had asked for details.

[158]Here Noether refers to Artin's paper [Art28b] in which a general structure theory of rings with double chain condition is given; today they are called Ar-

tinian rings. In this context the "double chain condition" is to be understood
as the simultaneous validity of the ascending and descending chain condition
for ideals. (At the time of this letter it was not yet known that the descending chain condition implies the ascending chain condition.) This is not to be
confused with Noether's ascending chain condition for ideals and descending
chain condition modulo every non-zero ideal, which Noether called "maximal
condition and restricted minimal condition" (this would lead to Dedekind
rings). Artin developed his theory for the arithmetic in maximal orders of
algebras [Art28a]. By the way, Artin's work provided the impetus for Hasse
to give a $\wp$-adic foundation for the entire theory: [Has31d]. Compare also
the letter of 25 June 1930.

[159]This refers to Dickson's book "Algebras and their number theory"
[Dic27].

[160]When Noether speaks of "my" passage, she refers to her paper "Hypercomplex quantities and representation theory" [Noe29]. There one can find
on page 667 the construction of the skew field K, over which a given "completely reducible two-sided simple ring with unit element" can be represented
as a full matrix ring.

[161]The lectures on "Algebra of hypercomplex quantities" in the winter
semester 1929/30 were written up by Deuring; these notes were later included
in the Collected Works of Emmy Noether [Noe83a].

[162]Noether refers to the second volume of Van der Waerden's "Modern Algebra" [vdW31], whose first edition appeared in 1931, Noether apparently
already had a proof copy, which she is sending to Hasse. However, the page
numbers given by Noether do not match with the page numbers of the printed
book; presumably the proof copy did not carry the final page numbers. In
the published version (first edition), at the given point there is a theorem
about maximal commutative subfields of matrix rings over a skew field D,
in particular that these are precisely the splitting fields of D. For splitting
fields of minimal degree this theorem was already given in the joint paper
of Noether's with Brauer [BN27]. Now Noether says that she did not know
the theorem which she finds in Van der Waerden in this generality. – This
passage of the letter gives evidence to what one reads on the title page of
Van der Waerden's "Modern Algebra", namely that this book was written
"using lectures" by Emmy Noether (and also by Artin). Even more: Emmy
Noether apparently took an active interest in the shaping of the book; we
infer from this letter that she thoroughly went through the proofs of Van der
Waerden's book.

[163]We were unable to determine whether Brauer presented this theorem
at the DMV meeting in Königsberg (Kaliningrad) in September 1930. Or
perhaps Noether knew that Brauer at the university of Königsberg had done
this in another lecture.

[164]This concerns a small conference on skew fields (today one would say "workshop") which Hasse planned together with Emmy Noether, and which was to take place in Marburg from 26 February to 1 March 1931. On the occasion of his appointment to Marburg, Hasse had been granted some money to organize such a workshop. He had in mind to gather all available experts to discuss his conjectures about algebras over number fields. (Regarding these conjectures see the two preceding letters from 29 and 24 December 1930.) — On 13 January 1931, Hasse gave a talk before the *Mathematische Gesellschaft* in Göttingen. The title was "On skew fields", and in the talk he introduced his conjectures on skew fields and algebras over number fields to the mathematical public, in particular his conjecture that every such skew field is cyclic. Very probably Hasse had talked to Emmy Noether about his plans for the "skew congress" on this occasion.

It seems unlikely that the idea for this congress goes back to a letter of Emmy Noether to Hasse from 1929, as Tobies suggests in a remark to a letter of Noether's to Paul Alexandroff. (Compare [Tob03].) In her letter of 13 October 1929 to Alexandroff she wrote: *"Hasse will go to Marburg as a successor to Hensel; I wrote to him about interrelated guest lectures (*zusammenhängende Gastvorlesungen*) but did not yet receive a reply."* But it is not clear whether these "interrelated guest lectures" were identical with Hasse's plans for his "skew congress". Some time before this letter to Alexandroff, Noether indeed had written to Hasse; see the letter of October 07, 1929. 3.18 In that letter she had asked Hasse whether Alexandroff could be nominated as successor of Hasse in Halle, but there is no mention of "interrelated guest lectures". It may be that Noether and Alexandroff had talked about a possible move of Alexandroff to a German university, and that the term "interrelated guest lectures" was used in this connection. But at the moment we do not know what Noether had in mind with this expression.

In this connection the following text from a letter of Hensel to Hasse seems to be of interest; the letter is dated on 21 October 1929. Hasse had informed Hensel about the fact that he had been granted a certain fund for guest lectures:

> *I am very glad that your request for a fund for guest lectures has been granted. This could become a good project of which the mathematics in Marburg and in Germany would profit considerably ...*

In particular we see that in those times not all German universities had regular colloquia with external scientists, as it is the case today. In Marburg such colloquia were installed only after Hasse took over in the year 1930. The impressive list of speakers in Marburg after 1930 includes the name of Emmy Noether, not only at the skew congress in 1931. (Compare the letter of 4 February 1933.)

[165]Shortly before this letter, Emmy Noether had delivered a guest lecture at the university in Halle. As we learn from this letter, Herbrand had come from Berlin to Halle for Noether's lecture.

[166]Emmy Noether refers to Hasse's class field report, which had appeared in three parts 1926, 1927, 1930, and also Hasse's papers on the theory of norm residues which finally lead to local class field theory. If Noether writes that Herbrand had learned the modern number theory from Hasse's papers, then this reflects the fact that the new developments in algebraic number theory were not yet widely known in France. Essentially it was Herbrand, Chevalley and Weil who, after their visits in Germany, prompted the quick development of modern algebraic number theory in France in the next generation.

[167]See [Her31]

3.26   Postcard of March 23, 1931

Principal ideal problem. Factor systems. Galois modules.

Lieber Herr Hasse!

Congratulations for the signing of the contract! I am looking forward to reading the matter in context soon![168]

The norm formulation of the principal ideal problem has now also become clear to me; I suspect that with consequent further development of the crossed products the solution will be available one day; but not today or tomorrow! I believe that one will treat much more general questions, and then it will become reasonable by itself.[169]

Your proof for the product theorem for factor systems contains all my deductions in another language! Instead of "automorphism ring" you say "elements commuting with the matrix units", make the preliminary remark on invariance (Artin, Dickson), where I appeal to operator isomorphism. For the proof that the rank is correct you count — even if not explicitly — the $e_{S,T} = e_S u_S u_T^{-1}$ (which I did not know), whereas I count the absolute components; everything else is literally identical! "Clarity" of a proof is indeed a relative term.[170]

From 26 March my address will be: Westerland-Sylt, Villa Richard.

Regarding the Galois modules one can, as Deuring told me, restrict to the center of the group ring since among the conjugacy classes the unit element forms its own class. This is the counterpart to the cyclic generation of the characters! Deuring has here posed himself the — probably difficult — question after the analogue of the Kummer field; that is, simple group, characters in the base field.[171]

Beste Grüße, Ihre Emmy Noether.

Notes

[168]This concerns the signing of a contract with the publisher Springer, according to which Hasse would (together with Hensel) write a book "Number Theory" in two volumes. The first volume would give an introduction to algebraic number theory based on Hensel's p-adics, and the second volume would treat class field theory among other matters. This plan was never fully realized. The manuscript of the first volume was completed in 1938 but didn't get published until 1949 when it appeared in the Akademie-Verlag [Has49a]. (English translation: [Has02].) The second volume was never written due to the rapid development of class field theory. In 1938 Hasse suggested Chevalley as a new author, but this didn't come to fruition.

[169]It is not entirely clear what Noether means by the "norm formulation of the principal ideal problem". The "principal ideal problem" asserts that every ideal of a number field becomes principal in its absolute class field. Perhaps by "norm formulation" she considers the transfer map (Verlagerung) of a group into a subgroup as a "norm", which indeed has a certain justification in that this transfer can be defined by means of a determinant, just as the norm in the usual sense is defined in an algebra. If this interpretation is correct, then Noether is looking for a hypercomplex proof of the transfer theorem. We remark that, at the time of the present postcard, the principal ideal problem had already been solved by Furtwängler [Fur29]. But this proof, although correct, was generally seen as unattractive; better proofs where sought. Let us quote the recollections of Olga Taussky, published in the book [TT81]. There she gives an account of a 1931 seminar with Emmy Noether. Regarding Furtwängler's proof Taussky writes:

> "... Once the proof of Furtwängler of the Hauptidealsatz came up and Emmy repeated what almost everybody said, namely, that it was an unattractive proof."

Compare also the letter from 8 November 1931, where Noether mentions a seminar in which "Frl. Taussky" also participates. This number-theoretic seminar seems to be the one recalled by Olga Taussky-Todd in the article mentioned above. That Noether is looking for a proof by means of crossed products can also be inferred from her letter of 19 December 1930 (last paragraph).

The much sought-after simplification of the principal ideal theorem was finally given by Iyanaga, who spent 1931/32 in Hamburg on a scholarship and attended lectures by Artin on class field theory. Iyanaga's proof was later published in the *Hamburger Abhandlungen* [Iya34]. In the preface to his paper he says that *"the greater part of the work is due to Artin"*. That proof found entry into textbooks through Zassenhaus, a doctoral student of Artin. His textbook [Zas37] was based on lectures by Artin. A further simplification, somewhat more in the direction of Noether's idea, was given by Witt [Wit54].

[170]Noether is referring to the proof of the product theorem she gave in her lectures. If A and B are central simple algebras over a field K, and if these are represented by factor systems $a_{\sigma,\tau}$ and $b_{\sigma,\tau}$ (with respect to a common Galois splitting field L/K), then the tensor product $A \otimes_K B$ is represented by the product $a_{\sigma,\tau} b_{\sigma,\tau}$ of the two factor systems. This theorem can be found in the notes for Noether's lectures made by Deuring, which were published posthumously. (In 1983 they were included in the collected works of Emmy Noether [Noe83b]). Deuring's notes did however circulate among interested mathematicians. Hasse was aware of them, so that Noether could refer to them in this letter without further explanation.

In the spring of 1931 Hasse was working on a manuscript in English which he wanted to publish in an American journal. It appeared in the Transactions of the AMS [Has32c]. This manuscript was finished in May 1931 and contains a detailed treatment of the theory of cyclic algebras over an algebraic number field. In the preface Hasse explains the reasons which caused him to publish in an American journal: *"These results do not seem to be as well known in America as they should be on account of their importance."* The paper contains among other things a thorough account of Noether's theory of factor systems which, as said above, were unpublished at that time. Noether had given her agreement for this. Apparently Hasse had told her about his proof of the product theorem and claimed that he had presented it more clearly; in this letter we find her response.

[171]Deuring apparently did not follow through with his ideas (but compare his paper "Applications of representations of groups by linear substitutions to Galois theory" in the *Mathematische Annalen* [Deu36]). The question of a characterization of Galois fields in generalization of Kummer's theory for abelian fields was taken up by Hasse in later years and considered in several

articles. (Compare "Invariant characterization of Galois fields" [Has50a], as well as the literature given there.) However, the center of the group ring is not sufficient there, as Deuring seems to have suggested.

3.27 Postcard of April 12, 1931

Cyclic algebras. Reciprocity Law. Artin's conductor.

Lieber Herr Hasse!

I have read your theorems with great enthusiasm, like an exciting novel; you have come very far![172]

Now I also wish the converse (it occurs to me that Deuring has wanted this for a long time): a direct hypercomplex foundation for the invariants, i.e., for the assignment of decomposition group and group of the non-commutative fields to the individual places; such that this is induced by a single assignment in the large; and thereby a hypercomplex proof of the reciprocity law! But this may take a while![173] At least you have, if I remember correctly, done the first step in your skew field paper with your exponents e_p?[174] Are the "crossed products" in English your invention? The term is nice.[175]

In the meantime I have had some further thoughts on the conductors; one has to apply the decomposition in Galois modules not only for Z but also for $Z \times \overline{Z} = \sum_S e_S \overline{Z}$; this directly yields the decomposition of the discriminant corresponding to the unit representations of the subgroups, in agreement with Artin. For the general Artin decomposition I still lack a few lemmas: one first obtains the decomposition into Lagrangeian root numbers; their ideal decomposition then must give the connection with Artin. I just do not know anything yet about the ideal decomposition in the general case![176]

Herzliche Grüße, Ihre Emmy Noether.

Notes

[172]This refers to the manuscript for the Göttinger Nachrichten [Has31c], where Hasse announced (without proof) all the essential results of his American paper [Has32c]. Compare also the previous postcard of 23 March 1931, as well as the following letter of 2 June 1931.

[173]A year later Hasse completed a manuscript in which he could fulfill Emmy Noether's wish. Hasse dedicated the paper to Emmy Noether on the occasion of her 50th birthday on 23 March 1932; the paper appeared in the *Mathematische Annalen* [Has33b].

[174]This refers to Hasse's paper on $\wp$-adic skew fields [Has31d] in the *Mathematische Annalen*. There Hasse constructed the local norm symbols with the help of Noether's theory of algebras, and this can indeed be regarded as the "first step" in the project proposed by Noether. But the ramification exponents e_p of local skew fields which Noether here mentions do not play a significant role for the local norm symbols. What Hasse had shown was that for a local central skewfield its ramification exponent e equals the index of the skew field.

[175]At that time Hasse was preparing the paper [Has32c] which he wished to publish in America in English language. For this he had received some linguistic assistance from the young American mathematician Engstrom who was a student of Ore in New Haven and spent the year 1930/31 in Göttingen with Emmy Noether. We do not know whether the English terminology "crossed products" for Noether's "verschränkte Produkte" is due to Hasse or was perhaps suggested by Engstrom. Since the publication of Hasse's American paper the "crossed products" were adopted in the English literature.

[176]Noether again refers to Artin's letter on what we today call the Artin conductors. Hasse had sent her this letter; compare Noether's letter of 10 October 1930. Noether apparently tries to obtain the Artin conductors by way of hypercomplex systems. Later, in August 1931, she sent a paper to Hasse in which she can obtain the Artin conductors with her methods at least in the tamely ramified case. Compare the letter of 22 August 1931.

3.28   Letter of June 02, 1931

Noether's comments to Hasse's American paper. Non-commutative foundation of class field theory and local-global principle. Engstrom. Herbrand.

Lieber Herr Hasse!

I have read your manuscript with great pleasure; everything looks so obvious — even the round-about way, as long as your definition of the norm residue symbol is not replaced by a better one — that one cannot see any trace of the piecewise finding of the proofs. [177]

The most interesting thing to me is the fundamental result (17.5.); I still believe — in spite of your skeptical letter of April — that the foundations of a hypercomplex set-up for class field theory lie here.[178]

However, I now think of stronger tools in this connection, namely a directly analogous treatment of ideal classes: the place of elements resp. classes (i.e., class of equivalent representations by D, D', ...) of $A = Z \otimes G$ would then be taken by the groupoid of ideals and ideal classes of A; the place of the assigned D the group of two-sided ideals resp. ideal classes; the place of the theorem of the inner automorphism the fact that two maximal orders (idempotents) are related by the transformation of an ideal and so forth. Then your objection that there is no hypercomplex proof of the quadratic reciprocity law would not apply any more; for one could call any proof building on genus theory as a hypercomplex one. In any case this may take some time, since nothing has been done yet! [179]

By the way, I find Theorem 2 almost equally beautiful; and I very much hope that in time you will manage to prove that everything can be generated cyclically. It does look very plausible that there should be no division algebra unramified over its center. Does Brandt not perhaps have a proof for this? I seem to remember that he once spoke about these questions.[180]

Now I shall also give constructive comments for the middle part: I believe that you have managed to make the matter palatable to both the Americans and the Germans, without sacrificing too much of the concepts.[181]

I have a few remarks which I made up during the reading — to be used or not:

The product theorem is better formulated:

$$\tilde{A} = (\tilde{a}\tilde{b}, \tilde{Z}); \quad \tilde{a} = ae, \tilde{b} = be.$$

The multiplication of the factor systems must be done in a *field* (for instance in the splitting field Z or in $\tilde{Z}$), and *not* in the ring $Z \times \overline{Z}$, where one finds $a \cdot \overline{Z}$. For the proof one only has to add on p. 41: $\tilde{u}_S = u_S e = e u_S$, which was proved long ago; therefore $\tilde{u}_S \tilde{u}_T = \tilde{u}_{ST} a_{S,T} \overline{b}_{S,T} = \tilde{u}_{ST} e a_{S,T} e \overline{b}_{S,T} e = \overline{u}_{ST} \overline{a}\overline{b}$. (This is also the way I had written it.)

Likewise one has to mention on p. 37 that the ideals $(A \times \overline{A})e^S$ as conjugates are all of *equal length*; or of *equal rank*; for only this implies that the Ae^S are *not simple*, the possibility of splitting of the n^2 matrix units, hence the passage to $e(A \times \overline{A})e$ (this corresponds to my preliminary remark on the automorphism ring $e(A \times \overline{A})e$.) The end is repeated in 15.

By the way, do you not want to, as e for idempotent, change the notation for the *unit element* on p. 19, or omit it entirely. I believe it is not needed anywhere.

p. 40, third formula line from below, an e was forgotten on the very right.

The rest is only some remarks on citations:

p. IV top; my lectures are to be cited as 1929/30; for the crossed products are entirely from 30. In Van der Waerden II one finds the theory of the splitting field, but not the crossed products.

Why do you not write on p. 21: (8.7.2) $\frac{c_S^T c_T}{c_{ST}}$; analogously for the a on p. 19: that would be a more symmetric ordering!

p. 23, second line from above, first ~ 1 is used; it would be good to refer to Definition I, 5.

p. 28, (c) becomes easier to understand if you immediately say that you want to reach (10.6); this theorem of the inner automorphism is incidentally also proved arbitrarily often in the literature:

in Van der Waerden II, in Brauer, according to my knowledge also in one of the most recent papers of Albert; a citation would seem the right thing to do.

p. 32. One can phrase (11.2) in such a way that the converse *always* holds: one only has to add that A is the algebra of *smallest* degree for which the embedding holds; in (11.3): "and if Z cannot be embedded into any A of smaller degree" (irreducible representation of Z by D). This is the way one finds it for instance in Van der Waerden (and in my lectures), which may potentially also need to be cited (incidentally (11.2), (11.3), (11.4) are in part independent from Ω being perfect, which you still require in your remarks; again to be taken from Van der Waerden or my lectures).

p. VI. At the end of your remarks to p. 36 one may later want to say that Brauer has *applications* of factor systems to the hyper-complex, while I build the *theory* in a hypercomplex way.

These are only suggestions of course: I cannot read things I know well without remarks.

Engstrom was very happy with your English apart from the re-orderings; hopefully you will also be happy with his existence theorem! He is in any case very enthusiastic about everything he has learned in Germany.[182]

I am sending Engstrom's manuscript to Deuring, who has been impatient for it for a long time; he is with Van der Waerden[183] for a few weeks as a replacement, since his own assistant Winter is still in America; but will be back for 1 July. In mid-June Herbrand also wants to come, who is currently with Artin.

Beste Grüße, Ihre Emmy Noether.

Notes

[177]Hasse had sent a copy of the finished manuscript of his American paper [Has32c] to Noether. (Compare the previous postcard of 12 April 1931.)

The "round-about way" mentioned by Noether is the manner of defining the norm residue symbol. At this point Hasse could not yet define the local norm residue symbol in a purely local way, but needed to use global constructions: The proof of the fundamental properties of the norm residue symbol is done using Artin's global reciprocity law. Hasse himself calls this way "round about" in the paper and poses the problem of finding a purely local definition.

[178]The "fundamental result (17.5)" in Hasse's American paper is the theorem that the Hasse invariant of a local division algebra is independent of the representation of the algebra as a cyclic crossed product. One year later, in his paper [Has33b] for Emmy Noether's 50th birthday, Hasse takes up Noether's idea stated in the present letter. There he says in the introduction:

> *"Emmy Noether remarked that this invariant theorem proved by me provides a direct, entirely local definition of the norm residue symbol. She thereby disproves the final sentence in my paper [8] in an unforeseen manner."*

Here [8] is precisely Hasse's American paper [Has33b]. The final sentence in [8], which Hasse now considers disproved, says:

> *"Even if the theory of the norm residue symbol should, at some time, be carried far enough to avoid that round-about way, the proof of Theorem 1, in the manner here developed will be preferable, I am sure, for reasons of brevity and simplicity."*

("Theorem 1" stated that a cyclic algebra over a number field is uniquely determined (up to similarity) by the system of its local invariants.)

[179]Here Noether refers to her favorite topic again, namely the development of a "hypercomplex" theory of ideals resp. ideal classes; today we would subsume this under the name of Galois cohomology.

[180]"Theorem 2" is essentially the local-global principle for cyclic algebras. It is interesting to see that Noether jumps in at precisely this point and remarks that one should be able to deduce that every division algebra can be generated cyclically — which was indeed done in the joint paper Brauer-Hasse-Noether the following year [BHN32]. — In the papers of Heinrich Brandt we have not found the proof which Noether seems to remember.

[181]The "middle part" in Hasse's paper [Has32c] contains a treatment of Noether's theory of factor systems. Noether had given the permission for her theory (which had not been published yet) to be included by Hasse in his American paper. Hasse treats Noether's theory in his own way, not as abstractly as Noether in her lectures. Noether comments positively on this in the present letter, but one also sees by her way of commenting that she is not

as enthusiastic regarding this part as for the other parts. This is apparently not only due to the following critical remarks on details, but also the entire set-up of the treatment, which is not abstract enough for Noether. Only the fact that Hasse is explicitly writing his paper for "the Americans", who are perhaps not as familiar with Noether's abstract way of thinking, reconciles Noether with Hasse's style. (And she then also admits that there may be "Germans" who are still not open to her abstract theories.) — Compare also Noether's remark on the American audience at her lectures in Princeton, letter of 6 March 1934.

At this point it is perhaps appropriate to add some remarks about the genesis of the theory of factor systems of algebras. Even though Hasse's 1932 "American paper" [Has32c] which Noether discusses here, was the first in which this theory was published, priority is clearly due to Emmy Noether, who lectured on this topic in the winter semester 1929/30. The lecture notes were written up by Deuring and were circulated among interested mathematicians. (The notes were later included in Noether's collected works [Noe83a].) Hasse had also received a copy of these notes (at least an early version), as the letters indicate. Hasse clearly states in [Has32c] that the theory of factor systems is due to Emmy Noether and was developed in her 1929 Göttingen lectures. The heading of the second chapter of Hasse's paper is "Emmy Noether's Theory of Crossed Products".

Noether's theory concerns factor systems which appear for *Galois* splitting fields, and which refer to the automorphisms of the Galois group. In her lectures [Noe83a] Noether calls these "small" factor systems, as they only depend on two parameters (from the Galois group). Previously in her lectures, Noether had treated arbitrary separable splitting fields, even ones that are not Galois. In this more general situation there are "big" factor systems, which depend on three parameters, and which analogously can be used to describe algebras. These "big" factor systems were first defined and systematically investigated by R. Brauer in [Bra26], as cited by Noether. In his subsequent work Brauer used these factor systems to prove structure theorems of what we today call the Brauer group. In this more general sense it can be said that the theory of factor systems for algebras goes back to Richard Brauer.

Nevertheless, Brauer's factor systems are "irrational", as Noether puts it in [Noe33b], i.e., the factors do not lie in the splitting field k but in its Galois hull. Noether later posed the problem of deriving "rational" determining data (i.e., contained in the splitting field itself and its conjugates), to her doctoral student Werner Vorbeck. But Vorbeck's dissertation [Vor35] was not completed before Noether was forced to leave Göttingen. Officially, Vorbeck got his doctorate in 1934 under F. K. Schmidt, after Emmy Noether had sent her report on the dissertation from Bryn Mawr.) Vorbeck's dissertation has never been published and had no influence on the later development.

Nowadays, as Noether had foreseen, her "small" factor systems are preferred, being easier to handle especially from the point of view of cohomology theory.

Independently of the development in Europe initiated by Brauer and Noether, beginnings of the theory of factor systems also emerged in the US within Dickson's circle. In the German edition [Dic27] (in section 34) Dickson considers division algebras (and more generally simple algebras) which contain a *Galois* maximal subfield; it is shown that such algebras can be represented as a crossed product. Dickson does not require algebras to be associative, however. Noether's associativity relations are hence missing; only in the case of Galois groups of rather special kind (two generators) are explicit conditions stated reflecting the associativity of the algebra. In [Dic28] Dickson reports that he can state such conditions for arbitrary *solvable* Galois groups, and this is elaborated on the example of groups of order pq. We may assume that these results of Dickson were known in Germany. (Hasse had critically reviewed the book [Dic27] for the "Jahrbuch", and likewise the paper [Dic28].) Later Dickson's student Albert showed in [Alb31c] that every simple algebra is similar (in the sense of the Brauer group) to a crossed product. Here Albert only works with associative algebras. In this case the associativity relations become trivial from the outset, and they are not mentioned there either. The question of constructing algebras by means of factor systems is not treated in [Alb31c].

At least these passages from the papers of Dickson and Albert are to be considered as first steps in the direction of a general theory of factor systems of algebras. Albert explains this situation from his point of view in a letter to Hasse of 26 November 1931:

> *I cannot quite see what you mean about the theory of crossed products. Did not Dickson really first consider them? As to the general associativity conditions I obtained them in 1929 from my matrix representation of any "normal division algebra of type R" in my paper "The structure of pure Riemann matrices with non-commutative multiplication algebras". The matrix representation (of any crossed product) is on page 31 (section 7) of the Rendiconti del Circolo Matematico di Palermo (vol. 55, 1931) reprint which I sent you. I never published the associativity conditions but they are immediate consequences of the matrix representations. I showed them to Professor Dickson in July 1929 when the above paper was completed but he did not think them important enough to be published.*

By this account it is due to Dickson that the theory of factor systems of algebras was not investigated further in the US. Dickson had discovered the cyclic algebras and described them in terms of the norm factor groups. But the step from the norm factor group of cyclic extensions to the second cohomology group of arbitrary Galois extensions was not taken by Dickson; it was initiated by Emmy Noether. Compare Noether's own statement in the letter of 12 November 1931, and also the third note there. Also compare the first note to the letter of 19 December 1930.

Observe that in these years, i.e., 1928–30, Noether's theory of factor systems was apparently not generally known in the US. This only changed with the paper of Hasse in the "Transactions of the American Mathematical Society" (and incidentally was refereed by Albert for the Transactions).

As the memoirs of MacLane [ML05] tell us, Albert also related the situation described above also to MacLane. MacLane writes:

> *"Dickson's student, A. Adrian Albert, once told me that he had generalized Dickson's cyclic algebras by replacing the cyclic group with an arbitrary finite group, thus essentially defining the crossed product algebras. However, on Dickson's advice, he did not publish the idea, so the crossed product algebras were first defined in Germany..."*

And for "Germany", MacLane lists the names of Helmut Hasse, Richard Brauer and Emmy Noether.

[182]Engstrom had seen through and corrected Hasse's paper regarding language. See the letter from April 12, 1931.3.27 The existence theorem mentioned here is needed in the proof that every algebra is cyclic. The theorem reads: *Given finitely many places of a number field K, and given for each place $\mathfrak{p}$ among these a natural number $n_{\mathfrak{p}}$, there exists a cyclic extension L/K whose degree n is the least common multiple of the $n_{\mathfrak{p}}$, and whose local degree at each of the given places $\mathfrak{p}$ is a multiple of $n_{\mathfrak{p}}$.* Hasse had apparently thought of a proof of this theorem, and had suggested to Engstrom to work it out. Engstrom agreed to this, but did not end up doing so. The theorem was later proved in a more general form by Grunwald (a doctoral student of Hasse). Wang (a doctoral student of Artin) discovered a mistake in Grunwald's theorem in 1948, but Hasse's existence theorem in the original form given above is not affected by this. Compare also [Roq05a],[BHN32].

[183]At that time Van der Waerden was ordinary professor at Leipzig University.

3.29   Postcard of August 22, 1931

Noether's Hensel note. Local integral bases for tamely ramified extensions. Artin's conductors. Deuring on class number 1 for imaginary quadratic fields.

Rantum on Sylt, 22 August 1931 (until the end of August)

Lieber Herr Hasse!

Here is my Hensel contribution! In the end I took something different from what I had originally intended, so that you and Hensel can have the pleasure of some $\mathfrak{P}$-adics.[184] — The references are very incomplete, as there is no literature in Rantum. Perhaps you have a Crelle assistant who can fill in the missing page numbers, otherwise I will do so in the corrections. And then he shall put a curly $\wp$ in the third reference, your skew field paper, I do not know how to write this! [185]

My interest in this matter is that in the simplest special case one can see the difficulties present by a direct development of conductor theory.[186] For I have reached precisely this point also in a general way. Incidentally, in Satz 7 I have used the decomposition a bit more generally than one finds it in the Zahlbericht (there the discriminant is divisible by *one* prime number only), and I first have to check whether it is, as I suspect, correct. Otherwise I shall restrict the hypothesis in the corrections.[187]

Will you come to Elster? I intend to join by old habit; and you are also one of the regulars.[188] Engstrom seems to be making progress with God's and F. K. Schmidt's help! [189]

In the meantime Deuring has proved that of the two conjectures, Riemann's and Gauss's — for a given class number there are only finitely many imaginary quadratic fields — at least one is correct; but he does not know which one.[190]

Beste Grüße, Ihre Emmy Noether.

Notes

[184]Noether was sending her manuscript for the anniversary volume of *Crelle's Journal* which was due to appear on the occasion of Kurt Hensel's 70th birthday. The volume was to be finished and presented to Hensel on 29 December 1931, the date of his birthday. The deadline for the submission of the contributions was 1 September 1931. Apparently in order to keep this deadline, Noether used her August holiday on the island of Sylt in the North Sea to write down the manuscript. As follows from her remarks, the manuscript is not quite ready for printing; some editorial work still has to be done. — The topic of this paper [Noe32b] is the existence of local integral normal bases in fields with tame ramification. This work became widely known. It is remarkable that Noether works with Hensel's local methods in this paper. Under the influence of Hasse who had described local algebras in [Has31d], she had apparently come to like Hensel's local methods and recognized their importance — very much in contrast to her other works, in which she continues Dedekind's tradition of ideal theory. It seems that Hasse had noticed this feature of Emmy Noether's present paper. In his speech at Hensel's birthday during the presentation of the dedication volume, Hasse said [Has32b]:

> ... *In particular you may consider it a very special and pleasant sign for the power and importance of your p-adic methods that even Emmy Noether, the purest follower of the old Dedekind tradition, in this volume presents a treatise working entirely with your p-adic methods...*

Noether's paper also contains the (probably) first proof for the existence of normal bases for an arbitrary Galois field extension (not necessarily integral normal bases), but only for an infinite base field. Another proof, including also finite base fields, was later given by Noether's student Max Deuring [Deu32].

[185]Noether refers to to Hasse's paper about $\wp$-adic skew fields in the *Mathematische Annalen* [Has31d]. Noether cites Hasse's theorem that every left/right ideal of a local maximal order of a simple algebra is a left/right principal ideal.

[186]Noether's motivation for this work was not only the existence of normal bases, but an algebra-theoretic road to the theory of Artin conductors. Noether had been informed by Hasse about Artin's conductor theory (compare Noether's letter from 10 October 1930). Here, Noether can only treat the case of a *tamely ramified* Galois extension. In this case the theorem of the integral normal bases gives a simple statement about the module-theoretic

structure of the ring of integral elements of the field extension: it is isomorphic as a Galois module to the integral group ring. This isomorphism is used to define certain "conductor ideals" with the aid of the decomposition of the group determinant. Noether cannot show in general, however, that these ideals are indeed the Artin conductors; she only succeeds in the case of a tamely ramified cyclic extension of $\mathbb{Q}$ of prime degree by reference to Hilbert (Zahlbericht). Only much later, in 1983, Fröhlich could show that Noether's definition indeed agrees with the Artin conductors [Frö83].

[187]It turned out that this was not necessary. In the final text Noether explains in a footnote that in her case Hilbert's hypotheses can be weakened.

[188]In September 1931 the annual meeting of the DMV took place in Bad Elster. Hasse came to Bad Elster together with his friend Harold Davenport with whom he had previously been on a lengthy road-trip through Europe.

[189]This concerns an attempt of Engstrom's to prove the existence theorem proposed to him by Hasse. Compare the letter from 2 June 1931.

[190]Deuring reported on this result at the 1932 International Congress of Mathematicians in Zürich and caused some sensation. His paper appeared in the *Mathematische Zeitschrift* [Deu33].

3.30   Postcard of August 24, 1931

Cyclic crossed products. Levitzky. Korselt. Herbrand's death.

Rantum (Sylt), 24 August 1931

Lieber Herr Hasse!

Many thanks for your letter! Of course I cannot answer your question either — I believe one needs to leave such a thing until one comes to it from another side[191] — but since you want to have new ideas, the following remarks by Levitzki may perhaps be useful.[192] Levitzki directly constructs, starting from the matrix units c_{ik}, "completely split crossed products" with cyclic group. Presumably, if the matter can be attacked directly at all, one has to relate the trace normal form to the factor systems; this normal form can be convenient. It is the following: Let us put:

$$d = c_{12} + c_{23} + \cdots + c_{n1} \; ; \quad e_i = c_{ii} \; ;$$

the the basis of the matrix rings is given by $e_i d^\lambda$ ($\lambda = 0, \ldots, n-1$). The relations for matrix units imply the further ones:

$$e_i d = d e_{i+1}; \quad d^n = 1;$$

thus if one puts:[193] $Z = e_1 P + \cdots + e_n P$; $G = \{d, d^2, \ldots, d^{n-1}, 1\}$, then one has the crossed split product: $Z \divideontimes G$,[194] where now the cyclic group G generates the permutations of the components of Z (such products have also occurred, in the case of partial splitting, in your American paper).

One can realize *every* group G correspondingly: Let the matrix units be denoted $c_{S,T}$, where S and T run through G independently; put again $c_{S,S} = e_S$. If one puts $d_T = \sum_{S \in G} c_{S,ST}$, then the d_T give a realization (with factor system one), and the matrix ring again becomes equal to $Z \divideontimes G$; that is, with $e_S d_T$ as basis. — Whether this will lead you to new approaches I naturally do not know.

Levitzki more generally considered such "normal products", where the two factors hypercomplex systems A and B; and only commutator rules hold: $ab = ba'$, where a' for a in A. This can become important for the question concerning the extension of groups.

Another remark is that, whether one cannot directly show with the Minkowski method that for skew fields the discriminant with respect to the center always has to be different from one. Everything becomes so much simpler here that the validity also for algebraic center — by taking norms — does not seem impossible. At least for rational center one should try this. The use of infinite places beside the finite ones does seem plausible! [195]

Concerning diophantine equations I shall present to you and other interested people in Elster a paper by Korselt; he has also explained his method to me in a letter, and Blumenthal wishes for "friendly refereeing"; but my wisdom is insufficient. Perhaps he will also be there himself; it concerns three binary forms.[196]

I cannot stop thinking about Herbrand's death. Did you hear that he died in a climbing accident in the French Alps at the end of July? His father wrote to me today with further details.[197]

Herzliche Grüße, Ihre Emmy Noether.

Notes

[191]We do not know Hasse's question which he had put to Emmy Noether. But since Noether's reply contains a method to construct cyclic crossed products, we assume that now Hasse is working towards a proof of the local-global principle for algebras. This is in accordance with a letter which Hasse had recently sent to Richard Brauer, on 27 July 1931. In that letter Hasse had written:

> I would like to inform you about the present state of the sole open question about the cyclicity of central simple algebras. In my opinion this question is now ripe for attack, and I would like to tell you my plans ...

Hasse continues that it is easy to construct, for a given central simple algebra, a cyclic field which is a splitting field locally everywhere. This reduces the problem to the local-global principle for algebras over number fields. Hasse sketches a plan how to achieve this perhaps with the help of his local-global principle for quadratic forms. For, if a_i is a basis of the given algebra A, then the trace determinant $tr(a_i a_j)$ defines a quadratic form. But Hasse does not know how to apply the local-global principle for this form to the present situation of an algebra A. He writes to Brauer:

> I would like now to put this question into your hands.

Since in the present letter from Noether she discusses bases and trace forms of algebras, we believe that Hasse had sent a similar letter to her, and here we have her answer. — Concerning Brauer, he had replied to Hasse on 3 August 1931. He wrote that he is not able at present to say anything about Hasse's question and would need some time to think about it. Relying on Hasse's creative power he adds:

> I hope to be able to understand the problem in time when you yourself have solved it.

[192]Levitzky had gotten his Ph.D. with Emmy Noether in 1929. His dissertation was published in the Göttinger Nachrichten [Lev29]. The paper

about normal products which Noether cites here, appeared in the Annals of Mathematics [Lev32].

[193]Noether usually uses P to denote the base field.

[194]Noether uses the sign "$Z \rtimes G$" to denote a crossed product; G is a group and Z a G-module.

[195]If we interpret Noether correctly then the "Minkowski method" means applying Minkowski's theorem about lattice points in convex bodies in order to obtain an estimate of the discriminant of an algebraic number field; the base field is supposed to be the rational number field, i.e., $\mathbb{Q}$. But the present case is concerned with skew fields. Even if it would be possible to extend "Minkowski's method" to skew fields, this would not solve the problem since here the discriminant is meant not over $\mathbb{Q}$ as base field but over the center. Obviously Noether is aware of this problem since she writes that one should at least try "'Minkowski's method" for skew fields with center $\mathbb{Q}$. Compare also the footnotes in the next letters, concerning "Siegel's proof".

In any case, at the time of this letter there did already exist methods from analytic number theory of skew fields which admitted the needed estimate for the discriminant over the center. This was contained in the dissertation of Käte Hey 1927 with Artin in Hamburg [Hey29]. There the ζ-function of a skew field over a number field had been defined. Comparing its poles and zeros with those of the ζ-function of the center yields in fact, that the skew field (if it is not trivial) has at least one ramification point, finite or infinite. But this was not directly mentioned in Hey's thesis. Moreover, it was later discovered that Hey's thesis contained some errors, which however could be corrected by Max Zorn [Zor33]. Only Zorn's note made it clear that the results of Hey and Zorn do imply the local-global principle for algebras. It is strange that this had not been observed earlier. Hey's dissertation was well known among the number theorists of the time, although it never appeared in a mathematical journal. The paper was printed as an independent publication and was distributed among scientific institutions, as required by the university at that time. Compare the article of Falko Lorenz on the dissertation of Käte Hey [Lor05].

[196]In the year 1936 there appeared a paper by Korselt in the *Mathematische Annalen* [Kor36]. There he determines for a numerically given elliptic curve a system of generators of its rational points. Perhaps this may be a new version of the paper mentioned here by Noether, but we could not decide this. Korselt was a teacher at a Gymnasium, and he had obtained his Ph.D. in the year 1904 with Hölder in Leipzig. In Hasse's legacy at Göttingen University there are 14 letters from Korselt, from the years 1928–1931.

[197]Jacques Herbrand had studied in Germany 1930/31 as a Rockefeller fellow, with Emmy Noether in Göttingen, Emil Artin in Hamburg, Helmut

Hasse in Marburg and John von Neumann in Berlin. Originally he was interested in mathematical logic, but then he also studied algebraic number theory and class field theory, which he had learned from Hasse's class field report. (Compare Noether's letter of 8 February 1931.) Herbrand had given essential contributions to the development of local and global class field theory. His results were very much appreciated, as was his friendly and open minded personality. His sudden death was a shock to everyone who knew him, also to Emmy Noether as we see from this letter. By the way, Hasse had been informed immediately after Herbrand's death. In Hasse's legacy in Göttingen there exists a letter from André Weil dated 4 August 1931, where Weil writes:

> *...I have to inform you about a very sad news, the death of Jacques Herbrand. who few days ago had a fatal accident while mountain climbing in the Dauphiné. I don't have to tell you which heavy loss his death implies for science and particularly for number theory ...*

See also [Roq14].

3.31   Letter of October 04, 1931

Correction to Noether's Hensel-note. Deuring. Siegel's proof for discriminants of skew fields?.

Lieber Herr Hasse!

I am sending you another remark to my Hensel note and ask you to forward it to the printer, since it seems to be too long for correction in the galley proof.

Deuring's example seems to be quite instructive; it reveals the limit of the hypercomplex formalities. Artin had conjectured that all results of his paper would follow if the discriminant could be represented as a group determinant — but now this is not the case if taken literally.[198]

In the general case there cannot exist a group determinant since the splitting of the conductor depends on the ordering of the composition series, the latter being determined by the field structure — this shows clearly the interplay between hypercomplex and number theory. Galois modules have to be interrelated with inertia- and ramification fields and -groups. But at present I only have beginning rudimentary ideas.[199]

As a consequence of one of the main theorems in my note — operator isomorphism of $K|k$ as a Galois module with the group ring over k — I had obtained Speiser's solution of Klein's problem of homogeneous forms.[200] Now Deuring drew another very interesting consequence: The operator isomorphy leads explicitly to the relation between subgroups and and intermediate fields (Main theorem of Galois theory, which at the same time gets a new proof that way).) The intermediate fields belong precisely to those ideals of the group ring which induce the trivial representation of the corresponding group — more precisely to the largest such ideal. I suspect that this may become important for number theory yet.[201]

Would you please send me some time Siegel's proof for the discriminant of non-commutative fields? Or has he indeed only proved that for $n > 1$ also $n^{n^2} > 1$?[202]

Notes

[198]Deuring had constructed an example of a non-maximal order whose discriminant can be represented as the square of a group determinant. Noether wishes to include this example as a footnote into her Hensel note [Noe32b].

[199]Today it is common knowledge that in arithmetic questions of Galois modules it is important to take into consideration the action of inertia- and ramification groups. It seems of interest that Emmy Noether mentions this already in the year 1931 .

[200]Here the "main theorem" is nothing else than the existence of normal bases for Galois extensions $K|k$ of fields. This implies that K as a k-Galois module induces the regular representation of the Galois group G. Hence for every k-irreducible representation of G there exists a corresponding G-submodule within K, at least if the characteristic does not divide the group order. This is essentially "Speiser's solution of forms", see [Spe16].

[201]Noether is citing Deuring's paper [Deu32]. This contains a proof of the existence of a normal basis by means of the theory of algebras. His proof works for all fields, finite or infinite, whereas Noether's proof in [Noe32b] works only for infinite fields. By the way, Deuring's paper also contains a new proof of the existence of a primitive element. The paper is published in the *Mathematische Annalen* and carries 17 November 1931 as the date of receipt.

[202]There exists an undated postcard from Siegel to Hasse which starts as follows:

> *Lieber Herr Hasse, today in the morning while traveling home*
> *I thought once more about your question on the discriminant of*
> *skew fields ...*

It seems that Siegel had been visiting Hasse in Marburg and that Hasse had talked with him about his question on the discriminant of skew fields. On this postcard Siegel proved that the discriminant of a skew field D, if taken over $\mathbb{Q}$, always has absolute value > 1. Hence, if $\mathbb{Q}$ is the center of D then there exists at least one ramified prime. If this proof would be correct then this would imply the local-global principle for central skew fields over $\mathbb{Q}$. At that time Hasse was working on the local-global principle for arbitrary central simple algebras over number fields. (Compare the preceding letter of 24 August. There Noether had also proposed such discriminant estimates.)

Siegel himself seems to have doubts about this proof since at the end of the postcard he writes: "*Where is the mistake?*" In his next letter to Hasse on 6 July 1931 Siegel writes: "*Concerning the discriminant it seems that I had not used the correct definition. Would you please inform me about the definition which you use?*" And further: "*I am sorry that I could not help you for the realization of your beautiful idea.*"

3.32   Postcard of October 23, 1931

Galley proofs for Noether's Hensel Note. Dedication to Hensel. Siegel.

Lieber Herr Hasse!

I am sending you the galley proofs for my Hensel note, with some bad conscience because of my many corrections.[203] But I have more rapidly published than I am used to. However I am not to be blamed for the $[G]_{\mathbf{op}}$ instead of $[G]_{\mathbf{o_p}}$; I hope this can still be corrected? Do you believe that (beginning at §3) my correction T^{-1} instead of $T-1$ will be understood? Otherwise please explain in more detail!

You may add to the dedication the words *"Herr"* or *"Geheimrat"* if you regard this as necessary; personally I prefer the way of the mathematicians. "Gordan means more", as Gordan had once said.[204]

Many thanks for your efforts to explain Siegel's proof. Myself, I couldn't do something with it either, although I tried to connect it with factor systems and splitting fields. The order of magnitude is plainly wrong if one tries to apply Minkowski directly, at least if the absolutely irreducible representations are of degree greater than one.[205]

Notes

[203]The Hensel-note [Noe32b] is Noether's contribution to the Crelle volume dedicated to Kurt Hensel on his 70th birthday. See the preceding letters.

[204]This concerns the text of the dedication to Hensel. Hasse had sent to Noether, as he had done to other mathematicians, a letter of invitation to contribute an article to the Hensel-volume of *Crelle's Journal*. There he had proposed that every article should carry a text like "dedicated to Herrn Geheimrat Hensel on the occasion of his 70th birthday". Obviously Noether did not like to mention the official title "Geheimrat" in a mathematical paper. In the final version of the publication, however, Hasse had changed his mind and there appeared, on the first page immediately after the title page, a general dedication text covering each contribution, as follows:

> *The articles in this volume are dedicated to Kurt Hensel, editor of this journal since 1902, by his students, friends and colleagues on the occasion of his 70th birthday on 29 December 1931.*

We observe that this text does not contain the official title "Geheimrat"; it seems that Hasse had accepted Noether's dictum that omission of the title means more. — The fact that Noether cites Gordan at this point can be explained since Gordan in Erlangen had been her supervisor for her doctorate.

[205]As to "Siegel's proof" compare the note to the preceding letter of 4 October. It appears that Hasse had thoroughly analyzed that proof and found that in principle it was not useful for his purpose. For, Siegel writes at about the same time, on 21 October, the following: *"Many thanks for your presentation of my unsuccessful proof"*.

3.33   Postcard of October 27, 1931

Abelian crossed products are cyclic. Non-Galois crossed products. Further corrections to Noether's Hensel-note.

Lieber Herr Hasse!

Congratulations on cyclicity![206] I suspect that for the general case one has to consider the analogue to crossed products for non-Galois splitting fields, since the subfields belonging to cyclic subgroups are kind of "partial" splitting fields. Until now nobody has wanted to do this.[207] On this occasion it comes to my mind whether I should not be cited in footnote 1). For, you rely in your American paper on my theorems which until now have been published only there. I wish you further success.[208]

As to my note[209], may I ask you to add the following to footnote 9) (since in footnote 8) I mention abstract fields): *"If k is of characteristic p with p dividing n then this is meant with respect to irreducible composition factors for Galois modules."* In the proof of Theorem 1 the text should be more precisely specified: *"If $\mathfrak{o}_\mathfrak{p}$ is fixed as base"*. (For the discriminant has the value one with respect to the place in the *rational* number field, hence also if $\mathfrak{o}_\mathfrak{p}$ is *not* maximal, then $[G]_{\mathfrak{o}_\mathfrak{p}}$ not maximal.) I should have started already with maximal $\mathfrak{o}_\mathfrak{p}$! In the hypercomplex situation it is tacitly assumed (i.e., it can be proved) maximality with respect to the *rational* base field; in the commutative case one says "integrally closed" and then this is evident. — The fact that $K_\mathfrak{p}|k_\mathfrak{p}$ is a direct sum of fields, is a consequence of the fact that the idempotents and components of the direct decomposition of $\mathfrak{O}_{\mathfrak{p}^i}$ converge; this I had in mind in theorem 5!

Herzl. Grüße, Ihre E.M.

Notes

[206]This shows that Hasse had sent to Noether a manuscript about "cyclicity". We do not know the text of this manuscript, and its content cannot be determined from Noether's postcard. However we know from other sources that Hasse's letter contained a proof that every *abelian representable* simple algebra A is cyclic, if the base field is an algebraic number field.

Let us briefly explain the terminology used at that time. A is called "cyclic" if A is isomorphic to a cyclic crossed product. A is called "cyclic representable" if A is similar to a cyclic crossed product, where similarity is meant in the sense of the Brauer group of algebras. Same for "abelian" or "solvable" instead of "cyclic". Hasse (and Noether) are working towards a proof of the local-global principle which implies that *every* simple algebra over an algebraic number field is cyclic. But as long as this is not proved then Hasse carefully distinguishes between those notions.

The manuscript which Hasse had sent to Noether can be considered as a step towards the final goal. Apparently the manuscript was meant for publication, for Noether wishes to be cited therein, as we learn from this postcard.

[207]Crossed products for non-Galois splitting fields are due to Brauer [Bra26]. Emmy Noether had included this theory in her 1927 lectures. Hasse had got a copy of Deuring's notes of those lectures [Noe83a]. Hence he was informed about the subject and therefore it was not necessary that Noether explain it to him. When Noether said that "nobody has wanted to do this", she means that nobody has wanted to consider crossed products of non-Galois splitting fields in the arithmetic setting over number fields. Hasse seems to have interpreted this as a suggestion to use factor systems of non-Galois fields for his problem. We know this because some time later he explained it in a letter to Richard Brauer of 7 November 1931. At that time Hasse already used the Sylow arguments of Brauer and hence could reduce the proof to the case when the group is a p-group. He wrote to Brauer:

> *I hope that now the proof can be successfully completed, when using essentially your theorems about crossed products for non-Galois splitting fields.*

Hasse's letter to Brauer has 8 pages. He presents in detail what he has done and how he believes the induction procedure of his proof can be completed. However Hasse himself was not able to do this at that time.

But two days later it turned out that all this effort was unnecessary, for Noether had found a simple and almost trivial argument (see Noether's next letters). Hasse explains this new situation quickly in a postcard dated 9 November to Brauer:

> *In the induction argument one should not start (as I clumsily did) with the field below, i.e., the group on top, but the other way round. I have gone to great troubles and did not find the simple idea of Emmy.*

In a later letter to Richard Brauer of November 16, 1931 Hasse mentioned once more his original clumsy proof, und he adds *"that I had essentially finished my proof when Noether's postcard with the simplification arrived"*.

[208]Since we do not know Hasse's manuscript we do not know its footnote 1) either. Perhaps this footnote cited Hasse's American paper [Has32c]. There he had published, with Noether's permission, her theory of crossed products which had not yet been published elsewhere.

[209]"My note" means Noether's Hensel-note which had been mentioned in her preceding letters. All of the modifications which Noether is asking for here have been taken into account in the published version.

3.34   Letter of November 08, 1931

Trivialization and generalization of Hasse's result. Deuring will write report on algebras.

Lieber Herr Hasse!

Many thanks for your postcard! Attached is a trivialization and generalization of your results.[210] I propose that you include this as an attachment to your note, since in fact the reduction theorem is implicitly contained in your last postcard where you remark that the factor system splits in regard to every cutout of the cyclic subgroup. But now this implies that even every *solvable* algebra is cyclic.[211] By the way, you may change the text as you like, and also change the notations to be compatible to yours. Also a text explaining the situation may be added. Alternatively I would agree to a separate publication, but then it would perhaps not be possible to include it into the Hensel volume?

I suspect that for the general case one needs new number theoretical results, besides the non-Galois crossed products. Of course one cannot predict anything detailed. What interests me more is the question of the hypercomplex interpretation of the Hilbert-Furtwängler proof. But of course this would require reworking of a sizeable part of class field theory.[212] In this connection I would like to ask you for a copy of the galley proofs of the American paper if this exists already.[213] At present I am conducting a number theoretic seminar: Local class field theory, Artin's conductors, principal ideal theorem (Ms. Taussky is here), etc. The existing proofs will be presented but it is my aim to understand them in the hypercomplex way.[214] I have thrown out the discussion of the associativity conditions from your proof for abelian algebras; this may be of use for the proof of the principal ideal theorem.[215]

The *"Zentralblatt"* plans to add a new book series with the title "Report on new advances" (or something similar). Deuring has taken over the report on hypercomplex theory, about 80–100 pages, to be finished by the end of May. I do not know when it will be published. In any case all the new results should be covered.[216]

Beste Grüße, Ihre Emmy Noether.

Notes

[210]Noether refers to Hasse's manuscript where he proved that over an algebraic number field every abelian representable simple algebra is cyclic. Hasse had sent this manuscript to Noether, as seen from the preceding postcard of 27 October 1932. Now Noether sends Hasse a manuscript in which his arguments are simplified, so much that Noether calls it "trivialization". Moreover it turns out that Noether's new induction arguments allow the generalization of Hasse's result from abelian algebras to solvable algebras. Hasse planned

to include his manuscript in the Crelle volume dedicated to Kurt Hensel. Out of concern that it might be too late to publish her manuscript in the Hensel volume, Noether proposes including her manuscript as an attachment to Hasse's. However the very next day Hasse was able to combine Noether's arguments with Richard Brauer's Sylow argument to complete the proof that *every* simple algebra over a number field is cyclic. Hasse quickly composed a new manuscript, jointly with Richard Brauer and Emmy Noether, containing the complete proof. It was included in the Hensel volume. See the following postcard of 10 November 1931.

[211]Noether uses a somewhat different terminology as Hasse; she writes "solvable algebra" where Hasse says "solvably representable algebra".

[212]The "Hilbert-Furtwängler theorem" is the local-global principle for norms of cyclic extensions of prime degree [Fur12]. The whole induction proof of Hasse (and Noether) is based on this theorem, as the first step in the induction procedure. We see that Noether is not satisfied with this situation. She wishes a new proof of this theorem which would provide a better understanding of the role of algebras in class field theory. Noether's wish was realized by Chevalley in the year 1933 [Che33b].

[213]The "American paper", which had been refereed by Albert, was published with some delay in 1932 [Has32c]. It was usual at that time to send the galley proofs as kind of "preprints" to interested colleagues. However, for the American paper Hasse never received the galley proofs, perhaps due to a mistake in the editorial office of the Transactions. Therefore there appeared in the same volume of the Transactions another publication of Hasse, an additional note containing numerous corrections for his original paper [Has32a].

[214]Compare the postcard of 23 March 1931.

[215]Noether refers again to her "trivialization and generalization" of Hasse's proof in the abelian case. As noted above that proof had become obsolete.

[216]This concerns the new series of publications *"Ergebnisse der Mathematik"* which just now had been started by Springer Verlag (who also published the new *"Zentralblatt für Mathematik"*). This series was meant to continue the tradition of the reports which in the past had been published by the German Mathematical Society (for instance, Hilbert's number theory report [Hil97] or Hasse's class field report [Has26a]). Emmy had been asked to write a report on hypercomplex systems. But she had declined and proposed her pet student Max Deuring for this project. Deuring was at that time 24 years of age. He delivered the volume under the title *"Algebren"* [Deu35a] which became one of the outstanding publications in this series.

The contract with the publisher about Deuring's book is contained in the archives of Springer Verlag; it had been signed on 19 September 1931.

Originally five printed sheets were planned, i.e., about 80 pages. But this limit was not kept by Deuring, nor was the time limit "until the end of May 1931". Finally, on 27 October 1934 Deuring submitted his manuscript with about ten printed sheets. But Emmy Noether, who actively had advised Deuring while writing the book, convinced Ferdinand Springer to accept the manuscript despite its larger extent.

As a curiosity we remark that at the same time as Deuring, i.e., 1931, A.A. Albert was asked to write, in the same series, a report on algebras. Albert agreed, but several months later he received a letter of Otto Neugebauer, the editor of the series, cancelling the agreement. Neugebauer wrote that he had not known that "algebras" and "hypercomplex systems" are the same thing, and the latter subject was already covered by Deuring. (But later Deuring used "*Algebren*" as the title of his book).

3.35 Attachment to Letter of November 08, 1931

Reduction theorem for splitting fields.[217]

This concerns simple algebras A, D with center k where k is an algebraic number field; (A) denotes the class of all algebras which are similar to A, i.e., all with the same automorphism field D (the same division algebra).

Reduction Theorem. *If (D) splits locally everywhere — i.e., full matrix ring locally everywhere — and if $K|k$ is a splitting field of (D) and L intermediate field such that $K|L$ cyclic of prime degree, then $L|k$ is a splitting field.*

Remark. *From number theory the proof uses only the Hilbert-Furtwängler norm theorem for prime degree; Hasse's norm theorem appears as a consequence.*

Proof. Let Λ be isomorphic to L; since (D) is splitting everywhere, so is (D_Λ). But (D_Λ) permits cyclic splitting field $K|L$ of prime degree, hence — using Hilbert's norm theorem — $(D_\Lambda) = (\Lambda)$, and therefore L is a splitting field (because splitting everywhere means here $u_S^\ell = a$ with a norm everywhere, hence norm of an element from Λ, hence $(D_\Lambda) = (\Lambda)$). $\qquad\square$

Corollary 1. *If* (D) *splits everywhere and admits a solvable splitting field* $K|k$ *then* $(D) = (k)$ *is a full matrix ring. In other words: if* (D) *is solvable class and if the factor system is everywhere associated to 1 then the factor system is absolutely associated to 1.*

For, if $K, L_1, L_2, \ldots L_{s-1}$ is a composition series of intermediate fields such that $L_{i+1}|L_i$ is cyclic of prime degree then the Reduction Theorem, if applied repeatedly, yields the assertion.

Special Case of the Corollary. *If* $Z|k$ *cyclic, then this is Hasse's norm theorem.* (If a is norm locally everywhere, then a is norm globally.)

Corollary 2. *If* (A) *is solvable class and* $(A) \neq (k)$ *then ramified, i.e., the discriminant with respect the center is* $\neq 1$. *(This is another version of corollary 1, using the fact that everywhere split is equivalent to unramfied, according to Hasse, skew fields.)*

Corollary 3. *If* (A) *has a solvable splitting field then also a cyclic one, hence every solvable class is cyclic. This follows in view of Hasse's reduction immediately from corollary 1. Observe that a solvable field* $K|k$ *remains solvable over* Z, *when* Z *an intermediate field.*

Corollary 4. *The question whether every class* (A) *is cyclic, is equivalent with the question whether for every class* (A) *there exists a minimal splitting field which is Galois.*

Proof. 1. If there exists a cyclic splitting field, then also a minimal splitting field which is cyclic. Here, "minimal" means that no proper subfield is a splitting field (in the cyclic case it follows from Hasse that there also exists one which is embeddable into the automorphism field). — 2. If all (A) have minimal Galois splitting fields then also the everywhere splitting (D), but then necessarily $(D) = k$. For, assume $K \neq k$ minimal splitting field and Galois; let $K|L$ cyclic of prime degree hence L proper subfield. Then $L|k$ splitting field because of Reduction Theorem, hence K not minimal. Contradiction. Therefore every (A) cyclic in view of Corollary 3. $\square$

Corollary 5. Assume (D) splits everywhere and K is a splitting field. Then also any intermediate field T for which $K|T$ is solvable, is a splitting field. *(Proof as for corollary 1.)*

Notes

[217]This attachment has been written with typewriter. As Noether says in her letter, she proposes to include this text in Hasse's paper for the Hensel volume. But it turned out that neither Hasse's paper nor this attachment was published; see the next letters.

3.36 Postcard of November 10, 1931

This is beautiful! The breakthrough: Every simple algebra over a number field is cyclic.

Lieber Herr Hasse!

This is beautiful! And quite unexpected for me,[218] although the last argument of Brauer (every prime divisor of the exponent appears also in the index) is quite trivial.[219] I had thought that more number theoretic arguments had to be used, of the kind that the decomposition groups are solvable. Now Corollary 5, which I did not think to be important, gives the whole theorem.[220]

By the way, I obtained my trivialization after reading the new Crelle papers by Brauer, of which he had sent me a copy (eight days ago).[221] Something else! Brauer's reduction to prime power index is not necessary. Let $N|K$ be a splitting field, p a prime number dividing the index, L the fixed field of the corresponding Sylow group, and (D) splitting everywhere. Then L too is a splitting field (Reduction Theorem), but not divisible by p any more. Contradiction![222]

Very soon I shall return your draft of the manuscript. The correction we could quickly settle independently.[223] Your American manuscript is not urgent.[224]

Beste Grüße, Ihre Emmy Noether.

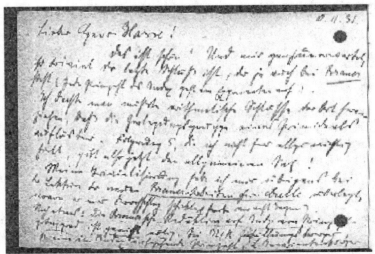

Original postcard of Nov 10, 1931 (back side now shown)

Notes

[218]This is the long-sought breakthrough. Hasse had written to Noether that he now has the full proof for cyclicity of all simple algebras over number fields. For this he had used, on the one hand, the induction argument which Noether had shown him in the preceding letter. On the other hand he used a Sylow argument which Richard Brauer had shown him shortly before, in a letter of 29 October 1931, and which reduced the general case to Galois algebras whose index is a prime power. The combination of the two methods yields the long-sought solution of the problem, together with Hasse's theorem which reduced it to the local-global theorem for algebras. In her last letter Noether had still believed that for the solution one would have to use non-Galois factor systems and, as she wrote, more results of number theory. But all that has now turned out to be unnecessary. This explains why the news came "quite unexpected" to her.

In the mathematical world the success of this project met with enthusiasm. For instance, when Artin heard of the proof of the Main Theorem he wrote to Hasse:

> ... You cannot imagine how ever so pleased I was about the proof, finally successful, for the cyclic systems. This is the great-

est advance in number theory of the last years. My heartfelt congratulations on your proof....

(This letter from Artin to Hasse is not dated but we have reason to believe that it was written around 11 November 1931.) Let us also cite from a letter of Siegel to Hasse, dated 11 December 1931, after Hasse had informed him about the final successful proof, which did not use any discriminant estimate about which they had corresponded earlier. Hasse had sent him a copy of the galley proofs of the paper which was to appear in the Hensel volume of *Crelle's Journal.* Siegel wrote:

> *...Many thanks for sending me the galley proofs. In fact, this is the best birthday present for Hensel, that his p-adic methods had led to such a triumphal result ... The pessimism which I feel against the future prospects of mathematics in general, is wavering again ...*

By the way, Siegel too had contributed a "birthday present" to Hensel. Siegel's contribution carries the title: "On the periods of elliptic functions" (Über die Perioden elliptischer Funktionen). It is the only paper of Siegel contained in *Crelle's Journal.*

[219]This theorem of Brauer is contained in his paper [Bra29]. Of course, Noether does not mean that this theorem is trivial. It is the *application* of Brauer's theorem in this context which she finds trivial. However, in the printed version of the Brauer-Hasse-Noether paper this theorem was not used any more but another theorem of Brauer was cited. (See below.)

[220]Here Emmy Noether refers to the attachment to her previous letter and there to the last "Corollary 5".

[221]As already mentioned, Brauer had written to Hasse on 29 October 1931. Together with his letter he had sent two manuscripts to be published in *Crelle's Journal* where Hasse was one of the editors. Now we see from Noether's letter that Brauer had also sent her copies of these manuscripts, and that she had read them immediately. Thereafter she had found the reduction method to the solvable case — what she calls "trivialization" — which she then had sent to Hasse. — Brauer's papers appeared in 1932 in *Crelle's Journal,* one in volume 166 [Bra32a], the other in volume 168 [Bra32b]. (The intermediate volume 167 was dedicated to Hensel, and Hasse had succeeded in getting there his joint paper with Brauer and Noether included.)

[222]Brauer's "reduction to prime power index" means the following theorem:

> *Any central division algebra of index m is the product of central division algebras of prime power index p^{r_i} according to the prime decomposition of the index: $m = \prod_i p^{r_i}$.*

This is contained in Brauer's paper [Bra29] which we have cited above. Noether's remark shows that this theorem is unnecessary for the present purpose. Accordingly, on 11 November 1931 Hasse wrote to Brauer informing him that Noether had *"removed your theorem of the direct decomposition"*. Noether's argument appears in the printed version.

[223]We see that Hasse had already attached to his letter a draft of the joint paper. Noether should check it and return it soon to Hasse who wished to include the paper in the Hensel-volume. (Compare the previous letter.) The Hensel-volume was to be completed on Hensel's 70th birthday on 29 December.

At the same time Hasse had sent a similar letter to Brauer. Again he attached a draft of the joint paper and wrote:

> *I am asking you to check the attached pages favorably and if possible quite quickly. As you see I have used the occasion to honor Hensel on his 70th birthday, and this is already on 29 December. We will launch a special edition (almost two volumes) and this paper should be included therein. This requires quick action.*

Indeed, Hasse managed to include the joint paper in the Hensel dedication volume, in time to present the volume to Hensel on his birthday.

[224]See the previous letter of November 8, 1931.

3.37   Letter of November 12, 1931

Noether's comments to Hasse's draft of their joint paper. Noether's methods. Albert.

Lieber Herr Hasse!

Now this is nice, and quite convenient for us that we have no further trouble with the text! But in my opinion you should mention in a footnote that it was you who edited the text — even if your style of writing is apparent. Also since we others are mentioned and you are not.[225] But concerning Albert you should weaken your remark. He has, in Theorem 19, only in the case of *cyclic* algebras shown that every prime divisor of the index appears also in the exponent. I am not able to find there anything of the general Brauer reduction.[226]

Then I would like to have some "historic" remarks added. For, in your reduction II you use, besides Brauer, essentially *my own* argument for closing the induction from above: If $K|\Lambda$ solvable, D splitting everywhere, then Λ is splitting field too — i.e., my corollary 5. Hence it seems adequate to add on page 2, *before* reduction II, the following text: "*... using the same argument which is also used in reduction* III *and which has been found* before *reduction* II." Or maybe with some different words. In this way it would also become clear that *you* have finalized the procedure, *not myself.* Otherwise it appears that I myself would have done the final step, using your argument in II, and both are not true.

Similarly I would like to be mentioned on page 4 in the 4th paragraph from below, or the "H. Hasse" there should be replaced by "we". For, already in the spring on our Hanstein walk I have told you that the factor systems are the correct generalization, when you told me that the norm-conjecture is false in the general abelian case. Perhaps you did not remember this and have later found out by yourself. Actually, I have told you this already in Nidden, in the form of the local principal genus theorem.[227], on the occasion of formulating the principal genus theorem for fields.[228]

Moreover I believe that you should mention yourself as the one who found Theorem 3 — I cannot sharpen it up and I believe you or Brauer will be able to do it.[229]

That's it for the "history". Now still some trifles: On page 5 the "reduced discriminant" is to be defined as "reduced norm of the different" (for $N\pi = p$; hence $N(\pi)^{n-1} = N(\mathfrak{d}) = p^{n-1}$ holds for the *reduced* norm). The non-reduced norm contains the same primes as the reduced discriminant but it does not coincide with it, nor with the unreduced discriminant.[230]

Don't you wish on page 1 (last paragraph, third line) to explain the notion "normal algebra" for German readers with the additional words "hence Ω as center".[231]

Typos: Page 2, first line replace II with III. Page 5, the remarks should be (5) and (6) (instead of (4) and (5)).

Of course I agree with the dedication text to Hensel. My methods are working and conceptual methods and therefore have penetrated everywhere.[232]

Notes

[225]This is the reaction of Noether to Hasse's text for the joint paper [BHN32]. We see that Hasse himself had finalized the text of their joint paper and already submitted this to *Crelle's Journal*, even before he had asked his co-authors Brauer and Noether to give their final consent. Certainly this unusual procedure was due to the rush in order to publish this paper still in the Hensel volume. As Noether had proposed, the published version contains a footnote saying: "*The editing of this note was done by H. Hasse*".

[226]Hasse had mentioned that a Sylow argument, as it was used in the paper, had been established earlier already by Brauer, and he had added:

> *Recently Albert had taken up this idea and he had developed for several general results of Brauer and Noether new proofs which are independent of representation theory.*

In this connection Hasse cited two papers by Albert in the Transactions of the American Mathematical Society [Alb31a], [Alb31b]. The "Theorem 19" which Noether mentions is contained in the second of these papers. Since she did not find anything of Brauer's Sylow theorems in this theorem 19 she proposes to weaken Hasse's text in this footnote. In her next letter she repeats this, but in the letter following the next (of 22 November 1931) she withdraws her criticism after Hasse had informed her about Albert's achievements.

[227]Hasse had included into his class field report (part 2) the Hilbert-Furtwängler theorem which claims the local-global principle for norms of cyclic extensions of prime degree. And he had conjectured that this holds generally for arbitrary abelian extensions of number fields. Later, in spring 1931 he had shown that this is true for cyclic, but not generally for abelian fields [Has31a]. It appears that he had informed Noether about this counterexample on a walk to the castle Hanstein near Göttingen. There Noether had told him that the generalization of the norm factor group in the cyclic case, is the group of factor systems in the case of arbitrary Galois extensions. — Nidden is a village on the Kurische Nehrung in East Prussia (today it is called Nida).

On the occasion of the meeting of DMV in Königsberg in the year 1930, Hasse and Noether had made an excursion to Nidden, and it seems that on this occasion she had told him that in the non-Galois case one should consider factor systems instead of norms.

[228]When Noether mentions the "principal genus theorem" then she uses the classical terminology for quadratic forms, or for divisors respectively. In those situations there exist genera and a principal genus. The classical principal genus theorem can be regarded as the vanishing of the 1-cohomology of certain groups of divisor classes or units. When Noether finds similar situations in any Galois module where the 1-cohomology vanishes then she tends to call it "principal genus theorem". The "principal genus theorem for fields" is nothing else than what today is known as "Hilbert's theorem 90", i.e., the vanishing of the 1-cohomology of the multiplicative group of a Galois extension. Hilbert had formulated this in his Zahlbericht [Hil97], but for norms of cyclic extensions only. Obviously, when Noether talked about this to Hasse in Nidden, she told him that for arbitrary Galois extensions one should work with factor systems instead of norms. This had already been observed by Speiser [Spe19]. — By the way, Hasse inserted the word "we" in the manuscript at the place Noether had asked for.

[229]This concerns the theorem that every irreducible matrix representation of a group of order n is realizable in the field of n^h roots of units, if h is sufficiently large. Hasse shows here that this is a simple consequence of the local-global principle. I. Schur had conjectured that this is possible for $h = 1$ already, therefore Hasse does not call it a "confirmation" of Schur's conjecture, but he says it is a step in that direction. Noether is of the opinion that Schur never conjectured that $h = 1$. But in the next letter she corrects this; therefore the manuscript got a footnote where the respective paper of Schur is cited [Sch06].

By the way, in the year 1949 Hasse published a short note in which he points out an "evident" error in his proof and corrects this [Has49b]. But in the same note he mentions that in the meantime Richard Brauer had obtained sharper results. Brauer had proved the full Schur conjecture, i.e., $h = 1$, and even more, namely that the field of e-th roots of unity is already sufficient where e is the exponent of the group. To this end Brauer had proved his famous theorem on induced characters [Bra47].

[230]For noncommutative algebras there exists a reduced norm and an unreduced norm; the latter is a power of the former. Similarly for the trace. Hence there are different notions for the different and the discriminant, according to which version of trace and norm are used. Following Noether, in the final version Hasse had inserted the word "reduced". However, he had set it into parentheses, obviously in order to point out that in the present situation it does not matter which version of the norm and of the different is used since both have the same prime divisors. Compare also the next letter.

[231]Hasse followed Noether's suggestion and included an explanation of the notion "normal algebra". — From Noether's proposal we learn that in the year 1931 this terminology, which was commonly used in America, was not yet well known in Germany. Later, after Deuring in his Algebra report [Deu35a] had used this terminology, this was accepted also in Europe. (But today one uses the more intuitive "central algebra" instead of "normal algebra".) By the way, in this paper Hasse uses the terminology "Algebra" for the first time, earlier he had used "hypercomplex system" instead, following Emmy Noether. Noether herself had never changed her terminology, in all her papers she uses "hypercomplex system".

[232]This sentence has become famous in the literature on Noether. It puts into evidence that she was very sure about the power and success of "her methods" which she describes quite to the point. But why did she write this sentence just here, while discussing the dedication text for Hensel? The answer which suggests itself is that, on the one hand, Noether wishes to express that, after all, "her methods" (as distinguished from Hensel's p-adic methods) were equally responsible for their success. On the other hand she does not care whether this is publicly acknowledged or not. In the present context "her methods" means two things: First, she insists that the classical representation theory be done within the framework of the abstract theory of algebras (or hypercomplex systems in her terminology), instead of matrix groups and semi-groups as Schur had started it. Secondly, she strongly proposes that the non-commutative theory of algebras should be used for a better understanding of commutative algebraic number theory, in particular class field theory.

3.38   Letter of November 14, 1931

More comments to their joint paper.

Lieber Herr Hasse!

I would like to reply by numbers.[233]

ad 2): I have looked at Albert again: In theorem 20 too only *cyclic* algebras are treated;[234] and again the theorem that every prime divisor of the index appears in the exponent. Also later in the text the assumption of cyclicity is kept. Therefore I propose to replace "this reduction" with "this index-exponent theorem in the cyclic special case". The citation of Albert's theorems should be canceled (maybe theorem 19 and 20 only); for in the later text (theorem 22 ff.) only Riemann matrices are discussed. Theorem 21 means only that $A \otimes A' \sim \Omega$ for cyclic algebras, where A' is inverse isomorphic to A. [235]

ad 3) und 1): I am surprised about the timing of the results. I believed that you, after reading my reduction, had finished the proof by using reduction 2, and this I wished to be cited. I could not believe that if one has the formally complicated reduction from K to L with $K|L$ solvable, one does not find the simple reduction when $K|L$ is cyclic; but sometimes such things happen. [236] Now I would propose that, according to your citation of Toeplitz, you would add to footnote 0 something like the following: that the proof is presented in the order as it had been found; a systematic proof would have to be written in the reverse order. As to myself, I did not understand your reversion of the systematic order, in contrast to "Toeplitz's pleasure". It is only because of the rush for publication that I abstain from "applying for systematization" and, instead, propose to change the text as said above. [237]

ad 5): It is absolutely correct that you put your name on the corollaries. I have formulated Theorem 2 only in connection with earlier information from you. — In the last sentence of Schur's first paper about the index he says: "*All known groups are representable by n-th roots of unity*". Since this is probably the only printed place, thus indeed it seems justified that this is called the "Schur hypothesis". I believe that R. Brauer would be able to give you more exact information. [238]

ad 6): Clearly Brauer has to be mentioned in connection with reduction 3. Indeed, it is a trivialization only, and it is absolutely trivial if one knows 2. [239] But perhaps in other situations the non-Galois crossed products will become important and will be needed, thanks to your computations! (compare 10).

ad 9): Already yesterday I observed that you meant the "ordinary" norm, but I did not write this to you since I — justly — guessed I would receive a letter from you today. The "reduced norm" leads to what Brandt has called the "basic number" (Grundzahl) in the case of quaternions; the reduced discriminant is the n-th power of this basic number.[240]

I am defining the "reduced norm of an ideal" locally. There every ideal is principal and therefore its reduced norm is simply the reduced norm of a generating element. The ordinary norm is the determinant of the transition matrix, and hence coincides locally with the non-reduced norm of a generating element, i.e., $N_{\mathrm{red}}(\mathfrak{p}_k) = N_{\mathrm{red\ in\ skew\ field}}(\mathfrak{p}_k) = p^r$ with $n = mr$; ordinary norm $N(\mathfrak{p}) = p^{rm}$.

ad 10): Congratulations! This looks quite promising. But I do not understand one argument in the proof! Why is the intersection $T_\mathfrak{p}$ of $U_\mathfrak{p}$ with $K_\mathfrak{p}$ the largest unramified subfield of $K_\mathfrak{p}$. After all, this is your contention: degree of $T_\mathfrak{p}$ is $f_\mathfrak{p}$, and this is the basis of all what follows. Would it not be possible that $K_\mathfrak{p}$ contains an unramified subfield which contains $U_\mathfrak{p}$ (then certainly it would be a splitting field), or alternatively an unramified field whose degree is prime to $m_\mathfrak{p}$ where $T_\mathfrak{p}$ is a proper subfield of $U_\mathfrak{p}$. Did I misunderstand something or is this a gap?[241]

In any case here is progress!

Herzliche Grüße, Ihre Emmy Noether.

Notes

[233]It appears that these numbers are taken from a letter of Hasse which he had sent to Noether as a reply to her last letter of 12 November 1931. We do not know exactly which topics were treated under the various numbers in Hasse's letter, but we can more-or-less figure it out from Noether's reply.

[234]In her preceding letter Noether had mentioned theorem 19 of Albert's paper.

[235]Concerning Albert see the next letter. By the way, Albert's theorem 21 is incorrect; he had formulated it for $A \otimes A$ instead of $A \otimes A'$. He corrected his error later in a one-page paper 'Erratum'. It appears that Noether had immediately discovered the error and formulated Albert's theorem correctly.

[236]Here Noether is concerned with the order in which the authors (Hasse, Brauer, Noether) have contributed to their joint paper. In his manuscript Hasse had divided the proof into three steps which he called reduction 1, 2 and 3. This concerns the main theorem that every simple algebra over a number field is cyclic. In reduction 1 Hasse reduced the proof to the local-global principle for algebras. In reduction 2 Brauer, with the help of a Sylow argument, reduced the local-global principle for algebras to the case that the algebra is solvably representable. Finally, in reduction 3 Noether reduced to the case where the algebra is cyclic. But for cyclic algebras Hasse had already proved the local-global principle in his American paper [Has32c] (this paper had not yet appeared but Brauer and Noether were informed about its content). Noether's astonishment that Hasse, after he knew Brauer's reduction step 2 did not immediately find reduction 3 but learned it from Noether's letter of 8 November only, is quite understandable. In any case, in retrospect the reductions 2 and 3 appear to us more or less obvious; Noether seems to feel this way too.

[237]We do not know the exact citation of Toeplitz which Hasse had given in his letter to Noether. Toeplitz in his papers had advocated the "genetic" method in mathematical education. This means that the theorems and theories should be taught in the order of their discovery and not (at least for beginners) in the logically systematic order. (Compare Toeplitz's textbook [Toe49], published in the yellow series of Springer Verlag.) We assume that Hasse, who was in friendly contact with Toeplitz, had used a citation of Toeplitz in order to justify his presentation of the proof of the main theorem, where he presented the steps of proof in the order how they were discovered. Actually, there was no time left for rewriting the manuscript, and this was accepted by Noether as we see in this letter.

[238]Compare the previous letter of 12 November 1931.

[239]Accordingly, the published text of the paper contains the sentence: *"By the way, R. Brauer too had found the third reduction independently of Noether."*

[240]Compare the preceding letter.

[241]We do not know the subject of point 10) in Hasse's letter. But from the notations which Noether uses, we suspect that Hasse had succeeded with the purely local definition of the norm symbol, or at least that he was on the way to this aim. This would be the first step to building local class field theory without the detour into the global theory. Compare letter of 12 April 1931.

3.39 Letter of November 22, 1931

Local and global class field theory. Albert again.

Lieber Herr Hasse!

Again I have to reply to several of your letters. Soon your results will arrive more quickly than we can grasp them.[242] By the way, the letters are already in the hands of Deuring (who is here for a couple of days since his father has passed away). Deuring wishes to include your results in his report.[243] Deuring is asking for a galley proof of your American paper. But there is no hurry. (Or do you not agree?)[244] Do you believe that the existence theorem too can be reached within half a year?[245] In any case: this is a breakthrough!

As far as I remember, at the end of my exposition[246] there is the following statement (but not yet on the pages which you have):

> The following facts can be regarded as the most essential part of local class field theory (more precisely: of the reverse theorem of local class field theory[247]):
>
> 1. (Hasse) The group $\mathfrak{A}$ of algebras over a $\mathfrak{p}$-adic base field whose index divides n is precisely the cyclic group of order n.
>
> 2. The subgroup $\mathfrak{A}_Z \subset \mathfrak{A}$ having Z as a splitting field where $Z|k_{\mathfrak{p}}$ is cyclic of degree n, coincides with $\mathfrak{A}$. [248]

So you see, in the local situation I had your concept, but I did *not* yet do the step into the global. The same concept — last year, when I did not yet have it fully — led to Deuring's reversion theorem,[249] i.e., that the group of algebras is invariant. But he also did not do the step into the global situation.

Earlier already, and recently more carefully, I have thought about another way to come from local to global, namely the extension of the Galois group G of $K|k$ with the group J of the absolute ideal classes of K (not of k),i.e., the purely multiplicative crossed product $J \rtimes G^{250}$ — because of the following: In this way one gets, perhaps only after a still stronger extension, to Schur's representation groups and their composition, i.e., to the "group of representation groups". In the $\mathfrak{p}$-adic case this will be finite again, in analogy to the group of algebras. In this way, a "similarity class" of algebras corresponds to a representation group ($\rightarrow$ corresponding division algebra) and its extensions ($\rightarrow$ matrix rings).[251]

For the local class field theory the above statement 1 is nothing else than Theorem 8, Part I of your report ($h_{\mathfrak{m}} \leq n$); here the use of analytic tools is not necessary because of the explicit description of the group $\mathfrak{A}$.[252]

Next week we shall discuss in our seminar your group of $\mathfrak{p}$-adic skew fields.[253] After that we shall discuss a sketch of the research report.[254]

It is good that you have settled the case with Albert. I had not looked at every issue and therefore missed some of his papers. Well, he seems to be quite able.[255]

Now I do agree with the footnote.[256] — Now everybody can find the proof, and this is a consequence of the fact that *you* had found it.[257] For the rest was trivial for everybody who was not bullheaded with this subject as you were.[258]

Herzliche Grüße, Ihre Emmy Noether.

Notes

[242]We do not know the contents of these letters of Hasse's. Noether's reply here gives some indication what Hasse had written to her. Obviously he had new results on the interrelation between class field theory and the theory of non-commutative algebras.

[243] "Deuring's report" is the volume on algebras in the newly started book series of Springer-Verlag called *"Ergebnisberichte"*. This has already been

mentioned in earlier letters of Noether. (Compare letter of 8 November 1931.) According to the publisher's agreement Deuring's book was to be finished in the spring of 1932. (But it turned out that it was delayed and it finally appeared in 1935.)

[244]Already in earlier letters Noether had asked for the galley proofs of Hasse's American paper. Since Hasse had not sent it yet she wonders whether perhaps he does not agree to include its result in Deuring's report. But Hasse never received the galley proofs for this paper; see letter of November 08, 1931.

[245]It is not clear which existence theorem Noether has in mind here. Probably she means the existence theorem for cyclic fields of given degree with prescribed local degrees at finitely many places. This theorem was used in their joint paper [BHN32] but its proof had not yet appeared. The proof was given by Hasse's Ph.D. student Grunwald in [Gru33]. There he stated an even stronger version of the existence theorem, namely that not only the degrees but even the cyclic field extensions can be prescribed at finitely many places. This stronger version, however, has turned out not to be true at 2-adic places. But the existence theorem for the degrees is nevertheless true. Compare, e.g., [LR03].

[246]Noether said "*Ausarbeitung*" which we have translated with "exposition". Noether has in mind Deuring's exposition of her lecture on hypercomplex algebra in the winter semester 1929/30 [Noe83a]. There indeed we find at the end the text mentioned by Noether here. Hasse had obtained a copy of this exposition. We learn from this letter that the text cited by Noether had not been among the pages which Hasse had received up to that time.

But Noether considers only *cyclic* extensions here (and in Deuring's exposition). Evidently she considers this as the essential part of class field theory. But the transition from cyclic to abelian was not trivial, at least not in the time of Noether's letter since the formal rules of algebraic cohomology were not yet generally developed. This transition was explicitly done by Chevalley [Che33a]. In Noether's lecture at the Zürich congress 1932 she admits that "new algebraic properties of factor systems" have to be developed, and she cites Chevalley.

As to the topic "local theory of algebras and class field theory", compare also the letter of 25 June 1930. There Noether had already mentioned this topic in her discussion of Hasse's paper [Has31d], still without knowing that the facts 1. and 2. could be obtained from the theory of algebras locally. But now Hasse had succeeded with this. That is the reason why Noether mentions this again.

[247]The local "reverse theorem" (*Umkehrsatz*) says the following: Every finite abelian extension of $k_{\mathfrak{p}}$ is class field in the sense that the index of the multiplicative group of norms in $k_{\mathfrak{p}}^{\times}$ equals the field degree.

[248]In these two statements, n is a fixed natural number and the algebras are Brauer equivalence classes.

[249]Noether refers to Deuring's paper [Deu31c].

[250]Noether uses the sign ⨯ for the crossed product.

[251]Noether's ideas in this paragraph are quite vague and we did not find a plausible explanation. It appears that she considered group extensions by means of factor systems in order to obtain global results which, e.g., were capable to explain the principal ideal theorem. (See also the last paragraph in Noether's next letter.)

[252]Noether compares the present results on local class field theory with the results in Hasse's report on global class field theory [Has26a]. (On this occasion we see that she has read Hasse's report thoroughly.) The inequality $h_\mathrm{m} \leq n$ was called the "first inequality" of class field theory; the notations were standard at the time and therefore Noether did not have to explain it. h_m denoted the order of the norm factor group in the global situation, here Noether compares it to the local norm factor group. n is the field degree. In the global case, the inequality could only be proved at that time with the help of ζ-functions and L-functions; therefore Noether pointed out that now, in the local case, analytic tools are no more necessary. (Of course, her desire was also in the global case to avoid analytic number theory, replacing it with the theory of non-commutative algebras.)

[253]This is the Brauer group of a 𝔭-adic number field. Hasse had shown that this group is isomorphic to $\mathbb{Q}/\mathbb{Z}$, by means of the Hasse-invariants. This is an immediate consequence of Hasse's American paper [Has32c] but there it is not yet explicitly stated. But the theorem is stated in Hasse's next paper [Has33b]. Perhaps the present exchange of letters with Noether occurred while Hasse was preparing that paper, and he informed her step by step about his results. This would explain the first two sentences of Noether in this letter.

[254]Probably Noether has in mind Deuring's report, which is mentioned several times in Noether's letters.

[255]This remark concerns a footnote which Hasse had inserted in their joint paper [BHN32]. This footnote describes Albert's contributions to the proof of the local global principle for algebras. In the two preceding letters Noether had wished to weaken this footnote since, in her opinion, Albert's achievements were not as important as Hasse had written. But now, after Hasse's remonstrance, she admits that she had overlooked one of Albert's papers. This paper is [Alb31a]. It seems of interest to see that it was Hasse who had insisted on keeping the footnote in its original form, despite Noether's objection. In the year 1931 Hasse had started an exchange of letters in which

both partners informed each other about their newest achievements. See also [Roq05b].

Remark: In [FS05] it is said that *"Emmy Noether seems to have served as the principal proponent of the footnote giving credit to Albert's work."* The present and the two preceding letters show clearly that this is not true. Originally Noether objected to the text proposed by Hasse, and she had changed her mind only after Hasse's remonstrations.

[256]In addition, Hasse had amended this footnote again. For, in the meantime he had obtained new information by Albert who, led by various letters from Hasse, had obtained the methods who led to the "reductions 1 and 2" of the joint paper of Brauer, Hasse and Noether [BHN32]. Although these results had meanwhile become unnecessary in view of the new proof in [BHN32], Hasse wrote that "Albert did contribute an independent part for the proof of the main theorem". (In order to make this explicit, Hasse and Albert published a joint paper where the contributions of Albert and those of Brauer-Hasse-Noether are shown in chronological order [AH32].)

[257]Originally Hasse had found a partial result only, namely the local global principle for *abelian representable* algebras over a number field. This can be seen from Noether's postcard of October 27, 1931 and her next letter. But meanwhile Hasse had succeeded in proving it for arbitrary algebras over a number field, although in a very complicated manner as he admitted. Although we do not know that proof, we know about its existence from a letter of Hasse to Brauer dated 16 November 1931. There Hasse wrote that he had achieved this just before he received Noether's postcard of November 10, 1931 with her "trivialization". It appears that Hasse had written the same also to Noether; in view of this she says here that Hasse had found the proof.

[258]Today we would see this in the same way. It seems curious that it took about a year until Noether's "trivialization" was found. This is mentioned again in Noether' letter of 27 January 1932.

3.40   Postcard of December 02, 1931

The boat picture.

Lieber Herr Hasse!

Would you please send me the negative of my boat picture (Danzig-Königsberg)? Just for some days? The people in Chicago are constructing a new building for their Mathematical Institute[259] — or they have already built it — and they wish to decorate their walls with pictures of mathematicians. Now your photo is the only acceptable existing picture of myself. Here I have only a quite battered enlargement of it (which had been used already for the negro woman), hence I would like to order a new copy for Chicago. As the photographer tells me it would be better to use the negative for this purpose. If mailing is inconvenient for you then I would ask you to enlarge the picture in Marburg; I will bear the expenses.[260]

Beste Grüße Ihre Emmy Noether.

The crossed product $J \gg G$ which I mentioned in the last letter has the following property: It transforms all those principal orders into each other which contain the principal order $\mathfrak{O}$ of K. (I have not examined the case $J \gg G \sim D \neq k$.)[261]

Notes

[259]This was Eckhart Hall.

[260]The boat picture had already been mentioned in the letter of 24 December 1930. — We do not know who had asked Noether to contribute a picture for the new Chicago institute. Perhaps it was Albert? But we know that she had sent the picture, and there it hung in Eckhart Hall for many years. It was last seen in the office of Herstein. Today it seems to be lost.

[261]Compare the letter of 22 November 1931. It appears that these ideas are connected with those in Noether's paper in the dedication volume for Herbrand [Noe34]. Compare the letter of 29 October 1932.

3.41 Letter of January 1932

List of names for reprints

January 1932[262]

Enclosed is your list and Brauer's addition.[263] I shall send my own list only after the distribution, for I would prefer to postpone the decision in this matter.

Isn't Bochner (18) identical with your Bochner (90)? The name of Ludwig (Dresden) seems dispensable to me; he did not work in group theory nor in number theory. Or do you have personal contacts? However, Kapferer in Freiburg should get a copy! He is lecturer (*Privatdozent*) there, and I know him personally.

Skolem is here at the moment. If the reprints would arrive within the next 14 days then I could personally give him a copy; otherwise please keep his name on your list!

In this winter semester I am lecturing hypercomplex, which gives me and the audience much pleasure. Shoda, a Japanese and student of Takagi, is here since fall. He is working well on these things; just now a paper on the group of automorphisms and rings of automorphisms will appear in the Annalen.[264]

Königsberg:[265]

1. Szegö

2. Reidemeister

3. Kaluza

4. Rogosinsky

5. Franz Meyer

Berlin:

1. A. Brauer

2. H. Hopf, presently in America, after return

3. Feigl

4. Hammerstein

5. Mises

6. Löwner

7. Mrs. Pollaczek-Geiringer

8. Remak

9. E. Rothe (Breslau)

10. E. Hopf

11. Fekete

12. Geppert

13. Buchner

Notes

[262]Noether did not give a precise date.

[263]These lists contain names of mathematicians to whom an offprint of the joint paper [BHN32] should be sent. Apparently Hasse is collecting the names from his two co-authors and will send them to the publisher who then will do the shipping. These names are of interest since they show with whom the authors had contact and exchanged offprints. However, Noether writes that her list is not yet finished. Moreover, Hasse's list is not preserved. Thus what we see below is Brauer's list only.

[264]The paper [Sho32] carries the title "On the Galois theory of semi-simple hypercomplex systems." If compared with Noether's own paper on this topic [Noe33b], Shoda's paper contains new proofs and also the generalization to semi-simple (instead of just simple) algebras. See also [Bra32a].

[265]This is the list of names which Noether had attached to her letter. More precisely: She had sent this list, and had written the letter on the margin of this list.

3.42   Postcard of January 27, 1932

Noether's version of the proof of local-global principle.

Lieber Herr Hasse!

Thanks for your new arrangement of the proof. But in my opinion you have complicated it, not simplified in view of what I had called trivialization, and what I wrote to you on a postcard after your Sylow-message. Things only became more complicated through your "historic" arrangement.[266] If one wants, the argument can be summarized as follows:

Suppose that p divides m [267]; consider the chain $k \subset L_0 \subset \cdots \subset L_f = K$. If K is a splitting field then so is L_{f-1}, hence also L_0; degree not divisible by p, contradiction. The new arrangement has the advantage that *without assuming splitting everywhere* to provide a little more information about the structure of splitting fields. Or do you strongly wish to avoid the induction argument?

I am sorry that my postcard which had been written in the evening (box cleared at $21^h 30$), did not arrive in time on Tuesday morning; the Institute will cover your expenses for the phone call.[268]

Beste Grüße, Ihre Emmy Noether.

Notes

[266]This concerns the arrangement of the proof of the local-global principle for algebras. Remember that Noether did not like Hasse's arrangement in their joint paper [BHN32]. (See letter of November 14, 1931. 3.38) We do not know the version which Hasse now has proposed but we see that again it does not satisfy Noether. Some time later, on the occasion of Noether's 50th birthday on 23 March 1931, Hasse dedicated his great paper on the structure of the Brauer group [Has33b] to Emmy Noether. In that paper he again included a proof of the local-global principle, and he arranged it precisely as Noether proposes here.

[267]Noether does not care to explain the notations; she assumes that the recipient (Hasse) would be able to figure out their meaning.

[268]At that time telephone calls out of town were quite expensive.

3.43 Letter of February 08, 1932

Structural proof of the extension theorem. Preparations for Artin's lectures in Göttingen.

Göttingen, 8.2.1932

For solvable groups it is possible to express " ~factor system One" by the factor systems of the composition factors, hence by norms; this may become important for the general conductors. [269]

Lieber Herr Hasse!

The structural proof which you are looking for was already contained in the exposition from which you extracted the crossed products, while you believed the first part. I am sending you the version which is ready for publication, although the applications — Galois theory and splitting fields — are not yet complete. For the theorem in question you should read §3 (p. 14) and the first two pages of §4, which are independent from the earlier §1 and §2; this concerns the "consequence of the extension theorem".[270] There I am considering only the case where K is non-commutative and normal over k, of finite or infinite rank. For, in principle the non-commutative case (K with center k) is to be put in the foreground since it is easier.[271] The case of a commutative K is treated in the §about splitting fields, simply by exchanging the roles, in which — notation as on page 18 — A is assumed to be of finite rank over P, and S a commutative finite extension field of P. This means the transition from A to $A_S = S_A$. The case of an infinite commutative Ω over P, i.e., A_Ω, follows either by extending with a splitting field, i.e., full matrix ring, hence without radical, hence A_Ω without radical, hence simple because the center is a field. Or alternatively, if Ω is algebraic over P then one shows that every two-sided ideal in A_Ω admits a basis in A_S already.

You see: If A is non-commutative and normal over P then it makes no difference whether the rank[272] is finite or infinite; but otherwise it does. The reason for this seems to be that the inseparable case is included; but in any case P is the precise field of invariants with respect to the group of inner automorphisms. Taking the two cases together shows that A_K remains simple, also when K not normal, the center of K of infinite degree, K infinite over its center.

I am sending also §1 and §2 which you recently rejected. I believe you will be interested to see how easily this can be used to obtain §4 and §5. Probably it will be sufficient for you to just check §1 and §2 without reading the details of proofs. Perhaps you will change your opinion.[273] Of course, the permutation theorems in §5 give essentially everything; these are the the theorems which R. Brauer has given in his Crelle paper; but since he worked in the commutative case only, he had to restrict himself to perfect fields.[274] Actually, I am not too interested in the generalization to imperfectness; my true interest lies in the non-commutative method.

These things are presented very briefly in Van der Waerden II. Several results, as e.g., part of the rank relations in §4, was not yet contained in the exposition[275], I have taken it from there[276] where it is formulated in a somewhat more special way. The proof which you are looking for you will also find, somewhat condensed, on page 175 in Van der Waerden – who uses one of my earlier lectures where, however, I did it somewhat complicated. Perhaps that is more convenient to you. As I just see, the extension to infinite in A_Ω is also treated there.

I have written to Mrs. Artin since here in the institute one has positive experiences with *her* letters; hopefully this works! Brandt writes today that he is not available on the 19th, but that does not apply to the 26th.[277] Probably I shall visit Halle in May.

Beste Grüße und bald auf Wiedersehen[278]

yours Emmy Noether.

The carbon copy I would like to have back some time.

Notes

[269]Apparently these lines were written after completion of the whole letter, between the date and the address, since there was no other free space on the page. This is an addition to the other text of this letter.

[270]Noether does not explain what she is writing about. But we have reason to believe that it is about the theorem: "*Let A be a simple algebra with center P, and let S be any extension field of P; then A_S is also simple.*" This theorem is contained as "consequence of the extension theorem" in §4 of Noether's paper on non-commutative algebra [Noe33b]. Among all papers of Noether on non-commutative systems this one is perhaps the most lucid (even though, or perhaps because it only clears up the basic concepts of the theory of algebras). This paper had not yet appeared at the time of this letter. As we learn from the letter, Noether had sent Hasse an extract of her manuscript. Of course, Hasse knew this theorem but he wished to have a "structural" proof, as Noether calls it. In particular the proof should include the cases when the field extension $S|P$ is infinite or inseparable. (Noether says "of second kind" instead of "inseparable".) The structure theorem which yields the required result is as follows: "*Every two-sided ideal of A_S is generated by a two-sided ideal of A*".

[271]Here we read in pure form Emmy Noether's dictum that the non-commutative algebra is simpler than the commutative algebra — if one considers the center as base field.

[272]Noether means the rank of S.

[273]We do not know why Hasse had "rejected" §1 and §2. Perhaps, without studying the details he had doubts because the results seemed him to be too general. When he refereed Noether's paper in the "Jahrbuch" he wrote only half of a sentence about these two sections: "*After a general introduction into the theory of automorphisms, modules and two-sided modules ...*" But then Hasse continued with a detailed review of Noether's paper, mentioning that the theory of direct and reverse representations is developed under quite general assumptions. And at the end of his review Hasse points out what Noether emphasized in this letter: that the results are proved "*without any restrictions on separability, contrary to all earlier foundations of the subject.*"

[274]Noether refers to Brauer's paper [Bra32a].

[275]Noether refers to the exposition, written up by G. Köthe, of her lecture in the summer of 1928. This lecture was the origin of Noether's paper which is discussed in the present letter. She mentions this exposition here because Hasse owned a copy and from it had learned her theory of factor systems, as she mentions in the letter. — By the way, the theory of factor systems is not contained in the published version [Noe33b]. Apparently Noether planned to deal with it in a second part, but this was not realized during her lifetime. The theory of factor systems was again presented in Noether's lecture 1929/30 where it was written up by Max Deuring. His exposition was published posthumously in the year 1983 [Noe83a]

[276]I.e., from Van der Waerden, Moderne Algebra II, §119 und §128 (1. Edition).

[277]Artin had been invited to give a series of lectures on the new class field theory in Göttingen. We see from this letter and the next postcard that Emmy Noether had proposed and organized these lectures which became an important event. It appears that Artin had not yet reacted to her inquiry; therefore Noether had written to Mrs. Artin. In the end these lectures took place from 29 February to 2 March 1932. — An exposition of Artin's lectures was written up by Olga Taussky and was widely distributed. A translation into English appeared 1978 in [Coh78].

[278]Hasse planned to visit Göttingen for Artin's lectures.

3.44   Postcard of February 11, 1932

Organization of Artin's lectures. Hilbert's theorem 90 for ideals.

Lieber Herr Hasse!

Just now I called Artin (or rather, had him called), his wife is still in Berlin. We have fixed 29 February, 1 and 2 March morning (i.e., Monday, Tuesday and Wednesday of the first week in the vacation period). Apparently Artin prefers this more than the weekend in between. Landau, whom I met by chance, was delighted, and Herglotz too told me today that he prefers these days after the end of the semester.

Now I hope that this will not be too annoying for you, perhaps Haussner will please you and move to the 26th or 27th.[279]

Probably I shall go to Halle on 11th and 20th. As Brandt has informed me, Krull will lecture twice, and Deuring about Artin's conductor. Myself, I will perhaps talk about about the connection between ideal classes in the commutative and in the non-commutative setting. Although at present these things are interpretations only (you say "words"), I believe that it will help to understand the situation. By the way, these considerations lead to a trivial corollary: If u_S, u_T, u_{ST} give factor system One after transformation, then they are symbolic $S - 1$, $T - 1$th powers (generalization of: If $N(\mathfrak{a}) = 1$ then $\mathfrak{a} = \mathfrak{b}^{S-1}$).[280]

Best greetings, yours Emmy Noether.

Notes

[279]Robert Haussner was professor in Jena. Perhaps Hasse had been invited by Haussner for a colloquium talk in Jena, or vice versa. But Hasse preferred to go to Göttingen for Artin's lectures.

[280]Noether had found that the group of ideals has trivial 1-cohomology.

3.45   Letter of March 15, 1932

Letter by Taussky, with Noether criticizing Hasse's article on Hilbert's papers on number theory.

Sehr verehrter Herr Professor[281]

Professor Miss Noether has told me that in your postface to Hilbert's number-theoretic papers some historic details seem to contradict each other, namely those which concern the Hilbert-Dedekind theory of Galois number fields. At one place the "completely new" discoveries of Hilbert are mentioned, and later it is reported about "former investigations" of Frobenius. (See the paper of Frobenius on the density theorems [Fro96] p. 689 and the paper and letters of Dedekind in his collected papers [Ded32] vol. I, p. 233 and vol. II, p. 414.) Since we will still get a revision, perhaps it will be possible to have small changes, e.g., the following:

I. On page 529, line 7 from below delete the word "completely" (völlig).

II. On page 533, in the paragraph after Artin's reciprocity law exchange the two sentences, and reformulate the first sentence somewhat like the following: "The existence and uniqueness of such a substitution S is contained in Hilbert's theory of Galois fields, which he developed independently of Dedekind and Frobenius."

Many thanks in advance and best greetings:

Olga Taussky.

Lieber Herr Hasse! [282]

I am acting as a historical troublemaker because of the Dedekind-volumes.[283] But this letter is not meant as official or even ultimative as it appears!

I was troubled over the fact that you represented the *connection of Galois theory and ideal theory* as being a *principally new* idea of Hilbert. But already in the year 1880 Dedekind at the cited place[284] had published at least as much that he could construct the Frobenius substitution. In fact, Dedekind had already established everything except the ramification groups (see the introduction to Frobenius). And the density theorem of 1880 (if published only in 1896) is just that connection![285]

Formally, this inaccuracy in your text leads to the contradiction mentioned above by Miss Taussky. I am well aware that these things became interesting for number theorists through Hilbert only — and that Hilbert may have recovered them independently.

Other people may be not such quibblers as myself, if you wish to leave everything as it is.[286]

Herzliche Grüße, Ihre Emmy Noether.

I have not made any progress with the principal genus theorem. First one would have to understand Brandt, in the sense of representation in domains of ideals.[287]

Notes

[281]This letter was written by Olga Taussky with typewriter, and Noether had added some comments in handwriting. — Taussky received her doctor's degree 1930 in Vienna with Furtwängler, with a dissertation on class field theory. 1931/32 she belonged to the Noether group in Göttingen. She had been assigned to assist Courant in the preparation of the first volume of the complete works of Hilbert. This volume contained Hilbert's papers on number theory [Hil32]. Her task was to check Hilbert's papers for mistakes and misprints, as well as to check the corrections. This includes checking Hasse's article which provided an overview of the number-theoretic papers of Hilbert. That article [Has32d] had been placed at the end of the first volume of Hilbert's papers.

[282]The following comments were added by Emmy Noether.

[283]Noether was editing the collected works of Dedekind [Ded32], in cooperation with Öystein Ore and Robert Fricke.

[284]See the preceding lines by Olga Taussky.

[285]It seems curious that Noether does not cite Dedekind's paper where he develops the Dedekind-Hilbert theory (with the exception of the ramification groups, as Noether admits). That paper is contained in volume 2 of Dedekind's collected papers [Ded32]. Perhaps Noether supposed that Hasse knew that paper, and her only aim was to point out that Dedekind had established the said results earlier than Hilbert.

[286]Hasse accepted the first proposal of Taussky, he deleted the word "völlig". As to the second proposal, Hasse reformulated the first sentence in the sense as proposed by Taussky, but he did not exchange the two sentences. That is, he first mentioned Hilbert and then Dedekind.

[287]It is not clear which papers of Brandt Noether has in mind here. The relevant papers of Brandt and Artin on the ideal theory in non-commutative

maximal orders had appeared some years before. Perhaps Noether is working on the cohomolgy of ideals and ideal classes, a topic which Noether will talk about at the IMU-conference in Zürich in the summer of 1932.

3.46   Letter of March 26, 1932

Thanks for Hasse's dedication paper on the occasion of Noether's birthday. Deuring. Brandt.

Lieber Herr Hasse!

What made you think of spoiling me so? I have been terribly happy![288]

Now I have finished reading your paper and can therefore really thank you for fun and seriousness! I have solved the $m_{\mu,\nu}$ syllable riddle by carefully following all breakpoints — and partly eaten! Also the rose garland from the black-red α-coalition had made a fitting impression, especially since it framed the "arithmetician", the highest praise you can hand out.[289]

But how you managed to do the hypercomplex receiprocity law in such a short time[290] is a mystery to me! Now I also have to make progress in order to earn the non-commutative dedication! But I believe that your paper will help me; because I already see some things more clearly! Above all, the role of the reciprocity law which forces the splitting of algebras (α, Z, S) at the primes $\mathfrak{p}$ of the unit class of the corresponding ray class group — that is, at the "divisor"-classes $k_{\mathfrak{n}}$ if $\mathfrak{n}$ is in the unit class and $\alpha = \mathfrak{n}N(\mathfrak{C})$ — while the splitting of the ramification primes of Z is by definition given by $\alpha \equiv 1 \mod \mathfrak{f}$.[291] In my opinion, it is from this point that one can proceed to a reasonable formulation! In any case, the theory now has gotten a completely different direction.[292]

Among the references I would like to have added the small note by Deuring in the *Göttinger Nachrichten*. For that was the first consideration in that direction (March 1930) and it has helped me much, although it has become essentially obsolete today, and he had published it quite late. To my surprise, I see now that he wrote at that time (30 March 1930) at the end of his proof: "*This theorem could play a role in the direct development of the norm residue theory — which will perhaps be necessary for a reasonable proof of the reciprocity law — for the commutation theorem* $(\frac{\alpha,\beta}{\mathfrak{p}}) = (\frac{\beta,\alpha}{\mathfrak{p}})$."[293]

Incidentally, Deuring is asking you whether he could have, for a few weeks, a carbon copy for his hypercomplex report?[294] (Holiday address: Göttingen, Hospitalstr. 3.) If not I would now the manuscript register only with Blumenthal[295] (isn't it intended for the Annalen?) — and send it to him somewhat later, this would do no harm for the publication. I will ask him about that.

As to Brandt, his talk in Kissingen should be mentioned.[296] Don't you want to write on page 15 more precisely for the infinite primes: For $\mathfrak{p}$ real, $e_{\mathfrak{p}} = 1$ or $= 2$, depending on $\mathfrak{p}$ real or complex, if $\mathfrak{p}$ real then always $e_{\mathfrak{p}} = 1$. I would be able to insert this.

By the way, your algebraic corollary p. 10 (2.5) is just a special case of that which I need for the general definition of the conductor. I formulate it in this way: "*Let* $k \subset L \subset K$, L *normal over* k *and* $K|L$ *cyclic. If* K *and* L *are splitting fields of* A, *then with suitable normalization a factor system for* K *may be composed of a factor system for* L *and of Ones.*" In your case I would normalize as follows: $u^m = \alpha w, w^s = 1, uw = wu$. I too have found this fact in Albert.[297]

And now thank you again! It seems that you have been the one who had informed the algebraists about my birthday.[298]

In herzlicher Freundschaft, Ihre Emmy Noether.

Please return the Algebren (for Fitting or ?). Also for Deuring.[299]

Notes

[288]Emmy Noether had her 50th birthday on 23 March 1932. Three days later, she thanked Hasse for his greetings and birthday gifts.

[289]We will probably never know the details of the "fun" in Hasse' birthday gift (perhaps a decorated birthday cake?). But we do know the "serious" part of the gift, for this was the manuscript of Hasse's paper which would appear in the *Mathematische Annalen* [Has33b], carrying the date of receipt as March 1932. The paper was entitled: "*The structure of the Brauer group of algebras over an algebraic number field. In particular a new foundation of the theory of the norm symbol and the proof of Artin's reciprocity law with non-commutative tools. Dedicated to Emmy Noether on the occasion of her 50th birthday on 23 March 1932.*" This work, together with Noether's lecture at the Zürich congress 1932 is to be seen as the historic climax of the Hasse-Noether cooperation, which started with her postcard from 4 October 1924 and then led to the structure theory of algebras over number fields and to class field theory. Noether immediately grasped the importance of Hasse's work. She had read the manuscript within three days of receipt and now sends Hasse her comments.

[290]Noether means the time since November 1931, i.e., since the completion of the joint paper [BHN32] on the local-global principle for algebras over number fields.

[291]Here Noether refers to a detail in Hasse's proof of the sum formula for the invariants of a simple algebra (Theorem (6.53) in [Has33b]). In that proof Artin's reciprocity law is not used in its full power but only its special case for cyclic fields of roots of unity. That can be proved in a rather elementary manner. Hasse's achievement is to construct such cyclic fields suitable for the present purpose. (Incidentally, he uses the same corollary of elementary number theory which Artin had used in his proof of the reciprocity law in order to apply the cross-over method of Chebotarev.) The full reciprocity law of Artin, at least its essential statement, can then be derived from the sum formula for algebras. This fact is the reason why Noether writes that Hasse had given the "hypercomplex reciprocity law". — It is not pure coincidence that she mentions just this detail in Hasse's proof. If she writes that α has to be defined such that $\alpha \equiv 1$ modulo the conductor $\mathfrak{f}$ then this seems like a discrete hint that Hasse does *not* say this at the relevant point in his proof; he only says that α has to be prime to the conductor, although the congruence condition is tacitly used later in the proof.

[292]The main features of this "completely different" direction are the following: First, Hasse could now define the local invariants of a simple algebra in a purely local way. Up to now this had only been possible with the help of global class field theory. The way to the local definition has been suggested to him by Emmy Noether, see letter of June 2, 1931.[3.28] Since Hasse's invariant is closely linked to the local norm residue symbol, it was possible to establish also the local class field theory by purely local means, without using the global reciprocity law (as has been necessary in former papers of Hasse's.)

Second, Hasse was now able to prove the global reciprocity law with the help of the sum formula for the invariants of an algebra. (See the preceding note.) This made one of Emmy's dreams come true. She had long claimed that the theory of non-commutative algebra is governed by quite simple rules, and that this should be used to approach the more complicated rules for the algebraic numbers.

In his paper, Hasse develops the whole theory *ab initio*, although the proofs of already known facts are not always repeated here, but he refers to the relevant literature. But the local-global principle for algebras is proved here again, this time in the manner desired by Noether. (See the letter of 14 November 1931.3.38) Overall, Hasse's paper is to be considered as a self-contained, great design.

[293]Noether had already mentioned this paper of Deuring [Deu31c]; compare the letter of November 22, 1931. Hasse took up Noether's proposal and cited Deuring's paper in his introduction.

[294]At that time Deuring was working on his report about the theory of algebras [Deu35a], which Noether had already mentioned in her earlier letter of 8 November 1932. He wished to include in this volume the latest results, hence he was interested in Hasse's manuscript. In fact, the main results of [Has33b] are contained in Deuring's report. — By the way, since here a "carbon copy" was mentioned, we conclude that Hasse's manuscript was typewritten. The times when manuscripts for publication were handwritten appear to have ended in 1932.

[295]Blumenthal was the managing editor of the *Mathematische Annalen*.

[296]Hasse included this reference in his bibliography [Bra28a].

[297]This is the theorem (in today's terminology) that the Brauer group of $L|k$ is injectively embedded, by inflation, in the Brauer group of $K|k$. Albert's paper mentioned by Noether is published 1932 in the American Journal of Mathematics [Alb32b]. Noether's remark shows that Albert's papers had been carefully read by her (and also by Hasse). In a letter to Richard Brauer dated 2 April 1932, Hasse states that he had laboriously distilled this fact from Albert's paper. Hasse is surprised that this theorem is a simple consequence of §3 of Brauer's old paper [Bra28b], as Brauer had told him. It seems that Noether had not known this either, for she too mentions Albert in this letter. In the published revision of his manuscript, Hasse added at the end of section (2.5) a reference to Brauer's result.

[298]Apart from the algebraists, nobody in Göttingen seems to have taken notice of Emmy Noether's 50th birthday. Olga Taussky, who was in Göttingen at that time, reports in her personal recollections of Emmy Noether [TT81] as follows:

> *Emmy had her 50th birthday in 1932 and told me about it. She had commented that nobody in Göttingen had taken notice of it. But then she added: "I suppose it is a sign that 50 does not mean old."*

And Olga continues:

> *Outside of Göttingen, Emmy was greatly appreciated in her country ... It was the academic year 1931–32 and she was at the height of her power and proud of her achievements, knowing that her ideas were now being accepted.*

The recognition of Emmy Noether "outside of Göttingen" is also manifested by the fact that she received in Leipzig 1932 the Ackermann-Teubner Memorial Prize to promote the Mathematical Sciences (jointly with Emil Artin). See the postcard of November 30, 1932.3.61 Moreover, she had been invited as one of the main speakers at the International Conference of Mathematicians in Zürich 1932 [Noe32a].

[299]These lines had been inserted afterwards with pencil. It appears that Noether had sent the galley proofs for Deuring's report "Algebren" [Deu35a], or parts of it, to Hasse, for his information. But others too should be informed, for instance Noether's student Fitting; therefore Noether is asking for return of the galley proofs.

3.47   Letter of April 05, 1932

Artin's letters.   Principal genus theorem.   Simplifications by Chevalley and Herbrand.

Breslau[300], Hobrechtufer 15 (until 17 April)

Lieber Herr Hasse!

Here, with the best of thanks, the Artin letters back; a little late because of travel and because I wanted to make notes of the main things, which was not a routine task but led to some thoughts.[301] I do not quite see it clearly yet; in particular the ramification groups are not cyclic any more (see last letter,[302] page 3 bottom). It seems that the composition of "algebras of ideals" [303] H (first

letter, first page) does not match with the Brauer group of algebras at the prime $\mathfrak{p}$. Instead, in the process of localization (at a
ramification point) the units will play an *essential* role, such that
the abelian commutator factor group appears. Perhaps a direct
definition of the generalized norm symbol $\left(\frac{\alpha_{\sigma,\tau},K}{\mathfrak{p}}\right)$ will be able to
exhibit the details of the situation.[304] That definition too will
control the abelian group at the ramification points, or what do
you think? Probably you will proceed further from here!

Incidentally, I do not consider it impossible that Artin's approach
in the cyclic case will be able to avoid analytic number theory,
but this too is a dream of the future![305] Deuring noticed that the
differences $h(z)$ appear everywhere (page 2); this is a consequence
of the fact that the a_σ form a crossed representation, hence $a_1 \sim 1$.
If one normalizes $a_1 = 1$ then the m_σ^1 vanish which is convenient.
I have not yet checked whether this is a triviality or belongs to the
basics of reciprocity – that is what Deuring thinks. Incidentally,
the crossed representation by the a_σ is the "transfer" of the σ
generated by the group H — the ideal algebra![306] Now I am
interested in what you see in these things.

It is very nice that the index calculations are becoming clearer
and clearer![307]

I have lent the correction of the American paper[308] to one of my
people, I believe Wichmann. Also Schwarz is interested in it; I
will send it to you occasionally. I told Deuring that he should
return his copy to you; hopefully he won't forget!

I do not yet see whether my principal genus theorem is more
powerful in the non-commutative case than Artin's, but I doubt
this.[309]

I believe that the new idea that is missing will have to use
operator-isomorphy instead of isomorphy of groups!

Herzliche Grüße, Ihre Emmy Noether.

Notes

[300]Emmy Noether visited her brother Fritz who was professor at the Technical University in Breslau (now Wroclaw).

[301]Hasse had had an exchange of letters with Artin in March 1932 about factor systems. This concerned the question of whether it would be possible, using the theory of factor systems, to approach class field theory for non-abelian Galois extensions. The Hasse estate contains five letters from Artin to Hasse from that time discussing this question; see [FLR14]. Apparently Hasse had sent those letters to Noether, which she is now returning. On the whole, the results fell short of their expectations. Artin had finally realized that a "completely new idea" would be needed in order to describe the decomposition of prime ideals in Galois extensions. With factor systems, Artin writes, one obtains simply the old method of applying class field theory for subfields over which the whole field is cyclic.

[302]Noether means the last letter from Artin.

[303]"Algebras of ideals" are classes of equivalent factor systems in the group of ideals. Things which are connected with factor systems were called "algebras" by Noether, since there was not yet an established terminology for those things. Today we would speak of the 2-cohomology of ideal groups, considered as Galois modules. — In the local case, where the ideals form an infinite cyclic group on which the decomposition group 3 acts trivially, the "algebras of ideals" are nothing else than the cohomology group $H^2(3, \mathbb{Z})$, and this is isomorphic to the character group of the commutator factor group of 3. This is well known today, but at that time Artin had to verify this by explicit computation. In the global case the situation is more complicated, and this is what Noether is worried about. — It is not easy to understand Emmy Noether's ideas which she tries to formulate in this paragraph. Obviously her aim is to find the "new ideas" which Artin had said are needed in the non-abelian case.

[304]Artin had proposed a definition of that "generalized norm symbol". However he remarks that this would lead essentially to the known Hasse invariants of algebras.

[305]For Hasse, avoiding analysis in class field theory was not an issue. He wished to develop methods which led to *lucid* proofs which are appropriate to the situation at hand. For him, this included methods of analytic number theory provided they met his criteria. – On the other hand, using the new index calculations by Chevalley and Herbrand (see below) it indeed became possible to avoid the use of analysis in class field theory. It seems remarkable that Noether already saw this at this stage.

[306]Here Noether is commenting on the result of a computation by Artin which today is known as "Lemma of Shapiro".

[307]This concerns the calculations for group indices in connection with the proof of the so-called second fundamental inequality of global class field theory. Hasse had mentioned, in his dedication paper for Noether [Has33b], that there exist "new simplifications of Chevalley and Herbrand", and for this he cites Chevalley's Paris thesis 1932. That was not yet published at the time of this letter, but its content was known among the specialists and was discussed in detail. E.g., Artin had written to Hasse in a letter dated 16 June 1931: *"I am excited about the new tremendous simplifications of class field theory by Herbrand and Chevalley."* Chevalley's thesis appeared later in the Journal of the Faculty of Science Tokyo [Che33b], together with a related paper by Hasse[Has34b].

It can be assumed that Noether's comment on those index calculations was a response to information she had received from Hasse. We remark that at the time of this letter the summer term 1932 had just started at German universities. In that term Hasse lectured in Marburg about class field theory. This lecture was worked out and then became widely known as Hasse's "Marburg lectures" [Has33d]. Hasse had planned this lecture according to the newest state of the art; in particular he had taken into account the new simple index calculations, and he had simplified them even more. It seems probable that he had mentioned this to Emmy Noether and that the remark in this letter is Noether's answer to it.

[308]Certainly these are not the galley proofs of Hasse's great American paper on cyclic algebras, for Hasse had never received the galley proofs for that paper. In consequence that paper contained many misprints. Therefore Hasse wrote another paper, published again in the Transactions of the AMS, where those misprints were corrected [Has32a]. It may be that the "corrections" which Noether mentions are the galley proofs of that second American paper, or a copy of Hasse's original manuscript for it.

[309]Noether had been invited as one of the main speakers at the International Conference of Mathematicians at Zürich, in September 1932. She intended to speak about her "hypercomplex" (i.e., cohomological) generalization of the principal genus theorem. It seems that she is already preparing for her Zürich lecture, and that this is the reason why the principal genus theorem appears more often in her letters of this period. The theorem is published in [Noe33a]. When Noether asks whether her theorem is "more powerful" than Artin's calculations (which were never published) then she is asking whether it leads to a stronger statement about the decomposition behavior of prime ideals in non-abelian Galois extensions. From today's perspective both, the principal genus theorem of Noether and the results of Artin's computations, are essentially equivalent to Hasse's local-global principle for simple algebras.

3.48   Letter of April 14, 1932

Noether's old symbolic computations. Noether's and Artin's conductors.

Breslau (from 18 April again in Göttingen),

Lieber Herr Hasse!

I am enclosing a letter in your capacity as Crelle editor. Does Crelle still accept such papers on invariants?

The author Petri was in the old days, – I believe 1904/05 – assistant to Jordan and to my father and a very capable person. He could well have completed his habilitation. But he became a teacher at Gymnasium, and he had published only one or two papers on algebraic functions, about 10 years ago, around volume 90 of the Math. Annalen. Klein was very interested in it. Apparently Petri has always worked on these old things.

The matter is related to my oldest Crelle paper (dissertation); between volume 130 and 140, about 137 or 135 or 136.[310] However I am unable to referee his paper; I have forgotten the symbolic computation with root and branch.[311] I believe Weitzenböck would be appropriate and perhaps also Van der Waerden who still knows the symbolism well. Perhaps one should send the whole thing to Weitzenböck since Petri is also considering submission to the academy. Crelle is only because of my old forgotten work.[312]

So please let me know and send the letter back in order to find a referee. I remember that Petri at that time made a strong impression on everybody; probably he was always a loner. Incidentally, once he had submitted another paper to Hensel, for Crelle, about fields of matrices, but later withdrawn for revision which was never seen again.

Did you fill your further vacations with literary activities as planned?[313] Or have factor systems etc. again provoked a strong stimulus for you? What has to be added in the sense of operator isomorphism, that is probably the analogue to Galois modules for ideals and ideal classes, and to Deuring's module of automorphisms. All intermediate fields belong to the latter, as Galois theory shows; I told you about that.[314] Perhaps something will arise out of here, if one is looking for the connection between my conductor belonging to the "*Hauptgeschlechtssatz*" and Artin's conductors; according to Deuring the latter are determined by his automorphism modules.

But enough of my phantasies! Viele Grüße, Ihre Emmy Noether.

Notes

[310]The dissertation (Ph.D. thesis) of Emmy Noether appeared in volume 134 of *Crelle's Journal* [Noe08].

[311]The first papers of Emmy Noether were concerned with symbolic computations in the sense of Paul Gordan.

[312]*Crelle's Journal* does not contain any paper by Petri. But in the year 1935 there appeared a paper by Petri in the *Sitzungsberichte* of the Bavarian Academy in Munich, with the title "On the discriminants of ternary forms" [Pet35]. That paper was refereed in the "Jahrbuch" by Weitzenböck, and in the "Zentralblatt" by Van der Waerden.

[313]Since Hasse had met Davenport in the year 1931, he had regularly been reading English literature which was recommended to him by the latter. Hasse also participated in English language literary circles. In a letter dated 15 November 1932 he reported to Davenport: "I am attending a course of Miss Diffené's on Translation and essay writing. It is quite amusing. We write essays on subjects suggested by A. Huxley's 'Those Barren Leaves'." Apparently Hasse had written to Emmy Noether about his English studies, and she is now inquiring about his progress.

[314] Perhaps Noether refers to the results of Deuring's paper [Deu36].

3.49   Letter of April 27, 1932

Köthe and local class field theory.   Hasse's Marburg lectures.
Chevalley.   Normal basis.

Stegemühlenweg 51 [315]

Lieber Herr Hasse!

I don't know whether you noticed that from Köthe's theorem right away follows the local class field theory for cyclic fields, i.e., the fact $(\alpha : \nu) = \prod_{\mathfrak{p}|\mathfrak{f}} n_{\mathfrak{p}}$.[316] And since you write about lecture[317] and Chevalley[318] I wished to inform you about it! In fact, with his invariance theorem[319] Koethe shows directly that in the local case the degree condition is also sufficient for splitting fields (for which you assumed the index relation $(\alpha_{\mathfrak{p}} : \nu_{\mathfrak{p}}) \leq n_{\mathfrak{p}}$ as known![320]) This means in particular: a field $K_{\mathfrak{P}}|k_{\mathfrak{p}}$ of degree $n_{\mathfrak{p}}$ generates the full cyclic group of algebras of order $n_{\mathfrak{p}}$; especially if $K_{\mathfrak{P}}|k_{\mathfrak{p}}$ is cyclic of degree $n_{\mathfrak{p}}$ then $n_{\mathfrak{p}}$ is also the order of the group $\alpha_{\mathfrak{p}}/\nu_{\mathfrak{p}}$; as was to be proved.

I wanted to write to Köthe that he should include this conclusion in his paper; he could refer to me. Or did you too told him already?

Your matrix description of Deuring's proof that came today is actually quite nice; if one considers $e^S = \sum z_i w_i^S = \sum z_i^S w_i$ with z isomorphic to the complementary basis for w, then this is the old proof, essentially at least.[321] But I assume that your presentation may be convenient for the identification of the hypercomplex defined conductor with Artin's conductor. Deuring has this in the case "without higher ramification" only.[322] My own approach of splitting without identifying, become clearer that way.

On the 6th I will probably give my talk in Halle.

Notes

[315]Emmy Noether had moved to a new address, therefore she is writing this new address in the head of the letter.

[316]Noether uses, like Hasse, the "group stenographic notation" in which groups are denoted with the same symbol as their elements. (See, e.g., Hasse's class field report, Part Ia [Has27a], or else his Marburg lectures [Has33d].) In the present case of a cyclic extension of number fields, α stands for the non-zero elements of the base field and ν for the norms. And $(\alpha : \nu)$ is the corresponding index of the norm group. Since the primes $\mathfrak{p}$ of the conductor $\mathfrak{f}$ are independent, one has $(\alpha : \nu) = \prod_{\mathfrak{p}|\mathfrak{f}}(\alpha_{\mathfrak{p}} : \nu_{\mathfrak{p}})$ where $\alpha_{\mathfrak{p}}$ denotes the group of non-zero elements in the $\mathfrak{p}$-adic completion of the base field and similarly $\nu_{\mathfrak{p}}$ the local norms. Hence Köthe's relation implies that $(\alpha_{\mathfrak{p}} : \nu_{\mathfrak{p}})$ equals the local degree $n_{\mathfrak{p}}$, for each prime dividing the conductor. This in turn is the main fact in local class field theory.

Noether cites Köthe's paper [Köt33]. This is contained in the same volume as Hasse's dedication paper for Emmy Noether but in a later part of this volume. At the time of this letter Köthe's paper had not yet appeared but Noether and Hasse were informed about its content. – Gottfried Köthe (1906–1989) had got his Ph.D. in Innsbruck and studied 1928/29 in Göttingen, mainly with Emmy Noether. Later he worked on functional analysis, under the influence of Toeplitz.

[317]In the summer term of 1932 Hasse in Marburg delivered a lecture on class field theory in its new form. The lecture notes were written by Hasse's assistant Franz and appeared as preprint [Has33d]. We had mentioned this already in our notes to the letter of 5 April 1932.

[318]Chevalley had given the method which could be used to extend local class field theory from the cyclic case to arbitrary abelian extensions. Hasse had contact to Chevalley, and it appears that he had informed Noether about it. Chevalley's paper [Che33a] appeared in *Crelle's Journal*; it was refereed in the *Zentralblatt* by Emmy Noether.

[319]The invariance theorem of Köthe says in the local case, that the Hasse invariant of a simple central algebra is multiplied by the field degree after extending the base field.

[320]Here Noether refers to Hasse's proof in the dedication paper [Has33b].

[321]In Hasse's diary there is a note, with the date of "April 1932", with the title "Existence of a regular basis for separable normal extensions. (Distilled from a hypercomplex proof by Deuring.)" Here, "regular basis" means the same as "normal basis". It concerns normal bases for field extensions, not

integral normal bases. Hasse's proof translates Deuring's hypercomplex proof [Deu32] into the language of matrices. Obviously, Hasse wanted for his lecture a proof which is independent of the theory of algebras. That proof in Hasse's diary is identical with the proof which appears in his lecture notes [Has33d] in connection with the index computations. It's probably this proof which Hasse had sent to Noether.

However, Hasse's proof works for infinite base fields only. But in that case, Noether herself had already given a proof which was independent of the theory of algebras, namely in her paper in the Crelle dedication volume for Hensel, where she showed the existence of a normal integral basis in case of tame ramification [Noe32b]. However, Noether later claimed that her proof there contains a gap. Indeed that proof is very short (three lines in a footnote). But Falko Lorenz had informed us that her proof can be worked out correctly. — Hasse's diary entry carries the remark which was added later only: *See Hensel, Crelle 103 (1887).* When we checked this reference we found that Hensel in the year 1887 had proved the existence of normal bases for finite fields, if considered as residue fields of number fields.

[322]In her paper [Noe32b] Emmy Noether had in case of a tamely ramified Galois extension defined, for every character χ of of the Galois group, a certain ideal Δ_χ. This definition was given with the help of the theory of algebras ("hypercomplex" as she says). She conjectured that these Δ_χ were identical with the conductors $\mathfrak{f}_\chi$ defined by Artin in [Art31]. But she could verify this in very special cases only. Many years later only Fröhlich could prove Noether's conjecture [Frö83]. If Noether says that Deuring could do that already then this may be a misunderstanding. In any case, Deuring had never published this.

3.50   Postcard of May 02, 1932

Köthe once more. Comparison with Hasse's proof. The new apartment.

Stegemühlenweg 51

Lieber Herr Hasse!

You have not misunderstood me at all; I only have, as I now realize, confused an old proof of mine with yours; I did not have the manuscript any more, that was sent to Blumenthal long ago.[323]

I had long since known that the relation $(\alpha_{\mathfrak{p}} : \nu_{\mathfrak{p}}) = n_{\mathfrak{p}}$[324] implies that locally every cyclic field is embeddable into a division algebra,[325] and that local class field theory can be hypercomplex formulated that way. Do you remember that I had asked you from Frankfurt,[326] whether you could prove with your methods — you had just sent me your skew field manuscript — $(\alpha_{\mathfrak{p}} : \nu_{\mathfrak{p}}) = n_{\mathfrak{p}}$?. But you could not do it at that time.

When I now received Köthe[327] the first thing that struck me was that now this old question was answered. I had not read that part in your manuscript or, more likely, I had thought about my old proof[328] and had only scanned that part.

Perhaps you will insert a brief note in the correction sheets at this point[329]; only now I understand a remark in one of your last letters that $(\alpha : \nu) = \prod n_{\mathfrak{p}}$ is hypercomplex done, and only the other index calculations remain. I didn't understand that remark! In any case, I have expanded my mathematical historical knowledge through this correspondence! [330] I will write to Blumenthal for quick corrections.

You should soon visit me in my new apartment: Math. Ann. vol. 1–105 and Crelle vol. 60–167 are available in the library guest room! [331]

Herzlicher, your Emmy Noether.

Notes

[323]Apparently, Hasse had objected to Noether's previous judgement in the postcard of 27 April 1932, that Köthe has first proven the local norm theorem. The result which Noether mentions and its proof was contained in Hasse's paper [Has33b] which he had sent her at the end of March. Now Noether admits that she had not read the relevant part in detail. On the other hand she writes that she herself had worked on this result since some time already.

[324]As to this group theoretical notation see the notes to the postcard of 27 April 1932.

[325]What is meant is: Every cyclic field extension of a $\mathfrak{p}$-adic field $K_\mathfrak{p}$ can be embedded as a maximal commutative field in a central division algebra over $K_\mathfrak{p}$.

[326]Remember that Noether had been in Frankfurt in the summer term 1930 as a visiting lecturer. See the postcard of June 25, 1930.3.20

[327]Noether means that she received Köthe's manuscript.

[328]There exists a letter of Emmy Noether to Chevalley, dated 12 December 1931, in which she writes that she has obtained this proof "*on the occasion of her seminar (local class field theory etc.)*" Moreover, she writes that "*it follows immediately from Hasse, Göttinger Nachrichten 1931,*.

[329]Indeed Hasse inserted in his paper [Has33b] (the dedication paper for Noether) a reference to Köthe at the relevant point, namely at theorem (2.5). It is interesting that Hasse formulates this reference as follows: "*For the proof see Hasse [BHN32] or better [Köt33].*" Here, [BHN32] is the joint paper about the local-global principle for algebras, whereas [Köt33] is just Köthe's paper in question; this appeared in the *Mathematische Annalen* immediately after the dedication paper [Has33b]. When looking through Köthe's paper it is noticeable that he often refers to the previous paper by Hasse, and he explicitly says that his paper is based on methods developed by Hasse. In a footnote he thanks Hasse "*for lots of advice on the drafting of this paper*". Thus Hasse had had a major influence on the preparation of Köthe's paper.

[330]Incidentally, in Hasse's Marburg lectures [Has33d] this norm theorem is *not* hypercomplex proved, i.e., not with the theory of algebras. Instead, the so-called "Herbrand Lemma" is used, which is concerned with the cohomology of cyclic groups [Roq14]. The theory of algebras does not appear in the Marburg lectures. This signifies that by now Noether's "non-commutative methods" in class field theory were gradually taken over by cohomology methods.

[331]Occasionally, Noether's seminars were held in this library guest room. Noether could not remain in this apartment for much longer than a year; thereafter she was forced by the Nazi government to emigrate.

3.51 Letter of June 03, 1932

On crossed products. Local class field theory. Chevalley's letter. Non-Galois splitting fields. Noether's Gauss reading.

Lieber Herr Hasse!

I am very pleased with your nice theorem, but above all with the fact that locally it gives the transition from cyclic to abelian. Perhaps this will clear up the role of the norms in the abelian case, which is not possible through factor systems.[332] Would you please tentatively send me the information from Chevalley? You will get it back soon![333]

I hadn't thought of the theorem myself, but I can prove and generalize it with crossed representations (page 190 of your American paper); with the u_S as a basis of the representation module.[334] In the same way I had Artin's proofs arranged for myself; didn't I write this to you from Breslau or at least indicate it? Incidentally, R. Brauer had told me years ago that it is possible to deduce $A^n \sim 1$ from the associativity relations (hence "first" observed by Artin is not correct); this can be found in one of his Crelle papers which just now have appeared.[335]

The crossed representation is simpler because the associativity relations are used in the trivial form only: $(u_S u_R)u_T = u_S(u_R u_T)$, in order to characterize the module as representation module. It is not necessary so use the more complicated form with factor systems. Everything else goes parallel to your proof.

Thus let $A = (KL \divideontimes \mathfrak{G}; a, b, c)$ with $\mathfrak{G} = \mathfrak{S} \times \mathfrak{T}$.[336]

I consider the crossed representations and "generalized" crossed representations with $\mathfrak{r} = (\ldots, u_R, \ldots)$ where R runs over all elements S in $\mathfrak{S}$:[337]

(1) $u_S u_{\overline{S}} = u_{S\overline{S}}\, a_{S,\overline{S}}$;

   $\mathfrak{r}u_S = \mathfrak{r}(a_{R,S})P_S$ where P_S means the permutation $R \to RS$.

   $\mathfrak{r}u_S u_{\overline{S}} = \mathfrak{r}a_{R,\overline{S}}P_{\overline{S}}(a_{R,S})P_S = \mathfrak{r}u_{S\overline{S}}a_{S,\overline{S}} = \mathfrak{r}(a_{R,S\overline{S}})P_{S\overline{S}}\, a_{S,\overline{S}}$;

   now passing on to determinants (here this is purely multiplicative, hence also possible in Artin's case where the $a_{S,\overline{S}}$ are ideals; Schur's transformation of the principal ideal condition);

$a_{S,\overline{S}}^n = A_{\overline{S}} A_S^{\overline{S}} / A_{S\overline{S}}$ with $A_S = \prod_R (a_{R,S})$

hence $a_{S,\overline{S}}^n = 1$.

(2) $u_S u_T = u_T u_S \, b_{S,T}$; (the inverses of your $b_{S,T}$, which is more convenient)

(α) $\mathfrak{r} u_T = u_T \mathfrak{r}(b_{R,T})$;

(β) $\mathfrak{r} u_T u_S = u_T \mathfrak{r}(b_{R,T}) u_S = u_T \mathfrak{r} u_S(b_{R,T}^S) = u_T \mathfrak{r}(a_{R,S})$
$P_S(b_{R,T}^S)$; on the other hand

(γ) $= \mathfrak{r} u_S u_T b_{S,T}^{-1} = \mathfrak{r}(a_{R,S}) P_S u_T b_{S,T}^{-1} = \mathfrak{r} u_T(b_{R,T})(a_{R,S}^T) P_S b_{S,T}^{-1}$;

now passing on to determinants (purely multiplicative):

$b_{S,T}^n = B_T^{1-S} A_S^{T-1}$ with $B_T = \prod_R b_{R,T}$.

(3) $u_T u_{T^*} = u_{TT^*} \, c_{T,T^*}$. Using 2)α) (and 1)) we get

$\mathfrak{r} u_T u_{T^*} = u_T \mathfrak{r}(b_{R,T}) u_{T^*} = u_T u_{T^*} \mathfrak{r}(b_{R,T^*})(b_{R,T}^{T^*})$

$= u_{TT^*} c_{T,T^*} \mathfrak{r}(b_{R,T^*})(b_{R,T}^{T^*}) = u_{TT^*} \mathfrak{r}(c_{T,T^*}^R)(b_{R,T^*})(b_{R,T}^{T^*})$;

on the other hand

$\mathfrak{r} u_{TT^*} c_{T,T^*} = u_{TT^*} \mathfrak{r}(b_{R,TT^*}) c_{T,T^*}$,

now passing on to determinants

$c_{T,T^*}^n = B_T B_{T^*} /, B_{TT^*} C_{T,T^*}$ with $C_{T,T^*} = \prod_R (c_{T,T^*}^R)$.

If one puts $A^n = (u_S, u_T, KL, , a^n, b^n, c^n)$ und further: $U_S = \overline{U}_S A_S$; $U_T = \overline{U}_T B_T$ then $A^n = (\overline{U}_S, \overline{U}_T, KL, 1, 1, C_{TT^*})$, as was to be shown.

This proof has the advantage that it generalizes almost without change to the case of a Galois subfield L of K (in place of KL).[338]

For, let L belong to the normal subgroup $\mathfrak{S}$ of $\mathfrak{G}$; let $\mathfrak{G} = \sum_T T\mathfrak{S} = \sum_T \mathfrak{S}T$ (with T representing the cosets); then the defining relations from (1), (2), (3) become

(1') $u_S u_{\overline{S}} = u_{S\overline{S}}$;

(2') $u_S u_T = u_T u_{S_T} b_{S,T}$ where $S_T = T^{-1}ST$;

(3′) $u_T u_{T^*} = u_{(TT^*)} u_{S,T^*}$ where (TT^*) is the representative of $TT^* \mathfrak{S}$,

hence $TT^* = (TT^*) S_{T,T^*}$.

(4′) $u_{TS} = u_T u_S$ (always possible normalization), hence by 3'):

$u_T u_{T^*} = u_{TT^*} c_{T,T^*}$!

Now part (1) of the proof remains unchanged, in part (2) one has to replace $b_{R,T}$ by $(b_{R,T}) P_T$ where P_T stands for the permutation $R \mapsto T^{-1} R T$; when passing to determinants this cancels out; thus part (2) remains too.

In part (3) one has first to say that for each element D from G the relations corresponding to (2′) can be computed:

$\overline{2}'$) $u_S u_D = u_D u_S b_{S,D}$;

one has to put $D = \overline{T}\,\overline{S}$ with suitable $\overline{T}$ and $\overline{S}$ which is unique; then $b_{S,D}$ is unique and one gets (since D can be used as a representative for $\overline{T}$):

$$b_{S,D}^n = B_D^{1-S} A_S^{D-1} \quad \text{with} \quad B_D = \prod_R b_{R,D}.$$

But now (3′) remains with $u_T u_{T^*} = u_{TT^*} = c_{T,T^*}$, up to additional terms which cancel when passing to determinants; hence one has indeed:

$$A^n = \left(L, \mathfrak{S}, \prod_R c_{T,T^*}^R\right);$$

since after introducing $U_S = \overline{U}_S A_A$, $U_T = \overline{U}_T B_T$, and generally $U_{TS} = \overline{U}_{TS} B_{TS}$ all $a_{S,\overline{S}}^n$ and $b_{S,D}^n$ with arbitrary D are changed into 1; this implies the splitting of the factor One (by general theorems which are not restricted to direct products; I use these theorems for the determination of the conductor, they appear already in Brauer's paper, I believe in *Mathematische Zeitschrift* volume 29). Of course, the new U_{TS} are in general not normalized any more for $T \neq E, S \neq E$, but this is irrelevant, although at first I got stuck with it.

I do not consider it impossible that on the basis of these general-
izations one can proceed from the cyclic local class field theory to
the solvable p-adic fields, hence to the behavior of all local Galois
fields — if only partially. Therefore I am interested how Chevalley
did it.[339]

Moreover, the simplification through crossed representations suggests to work also in group theory with crossed representations
instead with the complicated associativity relations, e.g., when
considering group extensions, or in connection with the principal
ideal theorem.[340] Maybe I'll get sometime someone who will do
this!

There is a student, not yet very independent, who wishes to become acquainted with these things; I have suggested to him to
give explicitly the crossed product in case of a non-Galois splitting
field using the embedding into the corresponding Galois splitting
field and splitting off the One-Algebra. I pretty much see how
this has to go, hope that he will hold out; others had quit, but
then I didn't know yet anything myself.[341]

Incidentally, while preparing my Zürich lecture[342] I have read
Gauss for once. A saying goes that any halfway educated mathematician knows the principal genus theorem of Gauss, but only
exceptional people know it in the realm of class field theory. I
don't know if that is true — my knowledge went in reverse order
— but at least I have learned a lot from Gauss with regard to
perception. Above all I learned that it is advisable to wait until the end proof before showing that factor systems lead to ray
classes. For, the transition of my version to Gauss' version is possible directly, independently from that. Only when specializing
to class field theory the conductor is needed. What I am doing is
generalizing the definition of genera by means of characters.

I hear that you are registered with the *Mathematische
Gesellschaft*.[343] So I am sure that you will come this summer.[344]
I have lectured at the beginning of May in Halle, afterwards I
was in Leipzig. Deuring had recently lectured in Erlangen, also
in Halle at the same time as myself.[345]

So goodbye and best regards!

yours Emmy Noether.

Notes

[332]In Hasse's diary [LR12] there is an entry, dated May 1932, entitled "A theorem on crossed products". Hasse's theorem there is identical with what Noether is proving below. So it can be assumed that Hasse had sent his theorem to Emmy Noether, and that now she returns her version of the proof. The theorem is about factor systems, and it has been used to build local class field theory. In the dedication paper [Has33b] Hasse had defined the norm symbol with the help of the invariant of an algebra, but for cyclic extensions only. With his norm symbol he could show in the cyclic case that the norm factor group is isomorphically mapped onto the Galois group. Now this is to be generalized to the case of arbitrary abelian extensions. This explains Noether's remark that "the role of the norms in the abelian case" will be cleared up.

[333]In the estate of Hasse there is a letter from Chevalley, dated May 12, 1932, in which the theorem in question is also stated. Chevalley writes: *"I have had the same idea about factor systems as you had."* Thus it appears that Hasse and Chevalley independently had found this theorem at the same time. Apparently Noether wishes to see Chevalley's proof too. The theorem is contained in Chevalley's paper in *Crelle's Journal* [Che33a] (as "Lemme 5"). By the way, Hasse in his diary mentions also a later entry, dated November 1934; there he cites a remarkable short proof by Witt. That proof appeared in *Crelle's Journal* too [Wit35b].

[334]The reference to Hasse's American paper [Has32c] means only that Noether uses the same methods as used there.

[335]This concerns the theorem that the n-th power of a factor system splits, where n is the order of the operator group G. This had been shown by Artin in a letter to Hasse (dated 9 March 1932), using explicit computation. Since Hasse had sent Artin's letters to Noether, she was informed about it. In the case of algebras this fact had indeed been known earlier already. But Artin's computations are valid for an arbitrary G-module; apparently this had been in the mind of Hasse when he wrote that Artin was the first who had observed it. – This example shows the increasing awareness of the mathematicians for the formal rules of cohomology. But it also shows the difficulties people had to overcome on the way to the proper notions and notations.

[336]In order to make the following computations understandable, let us give some explanations beforehand. Let $K|k, L|k$ two linearly disjoint Galois extensions of a common base field k (which is not explicitly mentioned by Noether). The Galois groups are denoted by $\mathfrak{S}, \mathfrak{T}$. The Galois group of KL is $\mathfrak{G} = \mathfrak{S} \times \mathfrak{T}$. Consider an algebra A which is given as crossed product of KL with $\mathfrak{G}$. The factor system is denoted by a, b, c where a consists of those factors a_{S,S^*} with $S, S^* \in \mathfrak{S}$, i.e., the restriction of the factor system to the subgroup $\mathfrak{S}$. Similarly c is the restriction to $\mathfrak{T}$, and b consists of the "mixed" factors $b_{S,T}$ and $b_{T,S}$ where $s \in \mathfrak{S}$ and $T \in \mathfrak{T}$. Let n be the order of $\mathfrak{S}$. The theorem to be proved says that the algebra A^n is similar to an algebra B which is the crossed product of L with $\mathfrak{T}$ and the factor system $c^{N\mathfrak{S}}$ where N is the norm operator from KL to L, i.e., $N\mathfrak{S} = \sum_{S \in \mathfrak{S}} S$.

This is the "Lemme 5" of Chevalley which we have already mentioned above. Chevalley used it in his paper [Che33a] to define Hasse's norm symbol for arbitrary abelian extensions of a local field, starting from Hasse's definition in the cyclic case [Has33b]. In this way the local class field theory could be transferred from the cyclic to the abelian case, as Noether had foreseen.

[337]Noether uses the symbol $\mathfrak{r}\ldots$ to denote a row of a matrix. Later these rows are combined to a matrix and then she takes the determinant of that matrix.

[338]This generalization is not correct; it appears that Hasse had told Noether; in the next postcard of 24 June 1932 she admits that she had made a wrong conclusion from the computations which follows here.

[339]Noether refers to the letter of Chevalley mentioned above.

[340]Every factor system defines a group extension. Perhaps this is what Noether had in mind. As to the principal ideal theorem of class field theory, this had already been verified by Furtwängler at the time of this letter [Fur29] on the basis of Artin's group theoretic reformulation of it [Art29]. A simpler proof was given by Iyanaga [Iya31]. Perhaps Noether is looking for another proof, more adapted to the situation.

[341]It is about Werner Vorbeck. He got his Ph.D. in Göttingen in the year 1935 with a thesis entitled "Non-Galois splitting fields of simple systems". At that time Emmy Noether had already been exiled by the Nazi government and was residing in Bryn Mawr (Pennsylvania, USA). Therefore F.K. Schmidt acted formally as Vorbeck's thesis advisor. He relied on a report which Noether had sent him from the USA. Although the thesis was found to be sufficient for a Ph.D., it appears that Noether was not very enthusiastic about it; apparently the candidate did not fully achieve the goal to her satisfaction. The thesis was never published in a mathematical journal. Non-Galois cohomology was almost forgotten until Amitsur introduced the so-called "Amitsur cohomology" and applied it to algebras.

[342]Emmy Noether had been invited to deliver one of the main lectures at the conference of the IMU (International Union of Mathematicians) in Zürich, September 1932. She lectured on "Hypercomplex systems in connection with commutative algebra and number theory." [Noe32a].

[343]Noether means the Mathematical Society in Göttingen.

[344]Indeed, in the summer 1932 Hasse lectured in Göttingen but not in the Mathematical Society there, but in Noether's seminar. Only later, in January 1933, he gave a talk at the Göttingen Mathematical Society.

[345]In Halle there lived Heinrich Brandt who invented what today is called the "Brandt groupoid". They serve to describe the ideal theory in maximal orders of a simple algebra. A report on Noether's lectures in Leipzig and her correspondence with Heinrich Brandt is contained in [Jen86]. — In Leipzig was Van der Waerden with whom Noether was in close contact. At that time Deuring was an assistant of Van der Waerden.

3.52   Letter of June 1932

Correction of proof given in letter of June 03, 1932. Determining the conductor.

Undated[346]

Lieber Herr Hasse!

I have indeed made a false conclusion at the place in question, leaving only a theorem of less interest that $A^n \sim (L, [?])$ (L inv. field of $\mathfrak{S}$), which in the case of direct products ($S_0 = E$) becomes your theorem ($\alpha_{S_0} = 1$, $\beta_{T_0}^{S_0} = \beta_{T_0 S_0}$).[347]

By the way, with the lemmas determining the conductor I meant precisely the one you gave, the "irrational" found at the beginning of §3 in Brauer, Untersuchungen ... I, Z[ei]tschr[ift] 28 (p. 688). There it is so simple that Brauer (688 below) only says: One recognizes without difficulty ... that is at least *almost* the same.[348]

With the conductors I need a somewhat more general version: I start only with the $a_{S,\overline{S}}$ equal to one; then I show that also the

$b_{S,T}$ from $u_S u_T = u_T u_S b_{S,T}$ can be made equal to one. Then we have your assumption. Moreover, this holds also if $\mathfrak{G}$ is not a normal subgroup. Then the crossed product with non-Galois L comes out, if K is the corresponding Galois field.

I know that my representation proof closely coincides with yours; however, I think it is likely that the considerations around the Principal Ideal Theorem become more understandable in this manner.[349]

About Artin's letter, I have considered that still much more remains to be obtained purely multiplicatively, for example, analogies to similar algebras, etc., but that is all unfinished.[350]

Herzliche Grüße, Ihre Emmy Noether

Notes

[346]The postcard is not dated, and the exact date cannot be determined from the postmark. The content concerns the correction of a claim Noether made in her letter from June 3, 1932. Therefore we have placed this postcard right after that letter.

[347]The formulas in this part of the postcard were difficult to decipher. In any case, it concerns the proof that Noether gave in her letter of June 3, 1932. The error occurs in the second part of that proof where Noether writes "*The proof has the advantage that it carries over almost without change to the general case where it deals with Galois subfields L of K (instead of KL).*" In this second part, an error occurred. The "theorem of less interest" that Noether mentions concerns the connections between the 2-cohomology of a finite group (given by factor systems) with the 2-cohomology of a normal subgroup. Noether calculated that the composition of the restriction map with the inflation map gives the n-th power map where n is the index of the normal subgroup.

[348]At the indicated place, Brauer [Bra28b] is dealing with the inflation map in the Brauer group.

[349]Here again the concern is about the lemma on factor systems, for which Noether gave a proof in the previous letter (first part) that she now calls "representation proof." Apparently Hasse had written that that was essentially the same proof as the one he had sent to her.

[350]By "purely multiplicatively" she apparently means the calculation with factor systems in (multiplicative) G-modules for a group G. Once again that is a step in the direction of the formalism of algebraic cohomology.

3.53   Letter of June 07, 1932

Further comments on Chevalley's work.

June 7, 1932

Lieber Herr Hasse!

Attached is Chevalley back.[351] Now I have a sufficiently old version; I requested a carbon copy from Chevalley, which he didn't have. I would like to have corrections[352] already for my winter lectures on non-commutative arithmetic, which first of all should bring the truly non-commutative but perhaps also a few commutative applications. I guess that also Deuring is interested in the corrections, for his report.[353] – For Lemma 2, page 8, [354] it would be good to cite Van der Waerden, last § 128? or R. Brauer; I believe in his Crelle paper which is still to appear.[355] Also, my now very delayed copy "Nichtkomm[utative] Algebra" states the same.[356]

Whether my generalization of the theorem on factor systems can do much for solvable fields, I now have doubts. In any case, one would have to proceed completely differently from Chevalley. For he uses here – different than in the proof of the lemma [357] – the two properties abelian and direct product fully, right from the beginning. Also the Inverse- and Isomorphism theorem don't hold without these assumptions! But perhaps the "sad remainder" can somehow be arranged such that it works not only for the largest abelian subfield.[358]

Viele Grüße, Ihre Emmy Noether.

Notes

[351]This concerns the letter from Chevalley to Hasse which Noether asked to borrow in her letter of June 3, 1932.

[352]She is referring to the correction sheets of Chevalley's paper [Che33a]. The work was intended for publication in Crelle's Journal and Hasse, as one of the editors, received a few copies of the correction sheets. As we do today with "preprints", at that time it was common to send correction sheets to interested parties so that they knew about it before the publication appeared.

[353]At that time, Deuring was working on his report on Algebras for the Springer-Verlag series *"Ergebnisse der Mathematik und ihrer Anwendungen"* [Deu35a].

[354]"Lemma 2" refers to Chevalley's "Lemme 2" in his paper mentioned above.

[355]In fact in Chevalley's publication Van der Waerden and Brauer were cited in connection with "Lemme 2".

[356]Noether is referring to her paper [Noe33b]. That paper is based on Noether's lectures 1928 and 1929, and there had been notes by Köthe and Deuring. Noether had planned to rewrite and to extend these notes but finally she did not do so; this explains the "delay" of the publication.

[357]Here Noether refers to Chevalley's proof of "Lemme 1" in his paper.

[358]Here again we see Noether's effort to extend local class field theory to non-abelian extensions.

3.54   Postcard of June 14, 1932

Emmy Noether prepares her Zürich lecture.   Chevalley's new Crelle paper and Noether's lecture manuscript from 1929/30. Mountain hike following the Zürich meeting.

June 14, 1932

Lieber Herr Hasse!

Can you answer a few trifles for the Zürich lecture?[359]

(1) Is this citation correct: *"Die Struktur der R. Brauerschen Algebrenklassen. Insbesondere: Hyperkomplexe Begründung des Artinschen Reziprozitätsgesetzes."* I have added Ann[alen] 104, which surely is correct. Do you already have the corrections? [360]

(2) What is the exact title of Chevalley? Is it *"La théorie des restes normiques"*? And do you already know in which Crelle volume it [will] appear? [361]

Moreover it would please me if you would make the reference to me in §1 more precise since my related comments in the lecture notes 29/30 have given the starting point to the [present] work. This is what Chevalley wrote today; I asked whether his definition of the norm residue symbol in [the] cyclic [case] is independent of the same-named [symbol] that I informed him about in my December letter. That was the case in fact, he had it already in October, but stimulated by my remark that locally the cyclic Isomorphism Theorem is expressed by the fact that every cyclic splitting field generates the entire group algebra of index a divisor of n. Perhaps Chevalley could also make that addition in the corrections. [362]

After Zürich don't we want to go somewhere in the heights together – I thought about M...[?] as a repeat of Nidden! [363]

I would like to spend 14 days already ahead of time in the mountains, perhaps also in Austria, where it is easier to transfer money. [364]

Beste Grüße, Ihre Emmy Noether.

Notes

[359]Noether is preparing for her lecture at the Zürich meeting of the International Mathematicians Union in September 1932. See [Noe32a].

[360]Noether is inquiring about the exact title of Hasse's article that she apparently wants to mention in her Zürich lelcture. It refers to the paper

that Hasse dedicated to her on her 50th birthday [Has33b]. See Letter from March 26, 1932. 3.46

[361]The exact title of the Chevalley paper is: "*La théorie du symbole de restes normiques*". It appeared in volume 169 of Crelle's Journal [Che33a].

[362]This refers to a citation that Noether wants to have included in Chevalley's paper [Che33a]. She wrote that to Hasse because she knew that Hasse carefully read all Crelle submissions, including Chevalley's, before releasing them to the printer. – At the end of the introduction of the printed version of Chevalley's paper, mention is indeed made of Noether's lecture notes.

[363]In connection with the 1930 meeting of naturalists and physicians in Königsberg, Hasse and Emmy Noether together undertook a side trip to Nidden on the Kurische Nehrung. Now she suggests making another side trip after the Zürich meeting of the IMU. (The name of the proposed destination M...cannot be read from the postcard.) But this trip didn't happen because Hasse had already made other arrangements..

[364]At that time there was a world wide economic crisis and it was difficult to transfer money from one country to another.

3.55   Postcard of June 16, 1932

Hasse invited to lecture in Göttingen. Lecture series of the Math Society in Göttingen. Chevalley again. Noether tries to understand Artin's letter.

Göttingen, June 16, 1932

Lieber Herr Hasse!

Concerning a lecture[365] apparently you were put on the list in error because I once said in winter that you would prefer to lecture about the joint note[366] in my seminar (and not in the Society) which you did, in connection with other things. But I hope that this misunderstanding will lead to an actual lecture in winter. In summer, as Neugebauer told me, already many guests are coming, especially since the math society is organizing a large lecture series (July 4–10) with Van der Waerden, Wegner, Walther, etc.   In winter Courant is here; the [knowledge of] latest number theory is always good for him and his overview.[367]

But as I said, you are on the list: in case you are interested in summer, register yourself, so that the Math Society gets to see the hypercomplex reciprocity law, and we can again be properly mathematical. Except for July 5, when Van der Waerden is here, my guest room is available [368]

But also my citation remark was misunderstandable.[369] I thought about including it in Chevalley's manuscript, which I assume you still have in your hands. But I appreciate that in your paper you have explicitly mentioned the notes of my 29/30 lectures.[370]

Chevalley writes at the end of his letter of June 13: "D'ailleurs tout découle naturellement de la remarque que vous faites à la dernière page de votre travail dactylographié sur la théorie des algèbres qui m'a donné l'idée de la définition en question (the N.R.S.[371]) and I would like to see that cited in spirit, not necessarily word for word. In case the manuscript is already gone, I request that Chevalley insert it in the corrections.

I'm trying once again to "understand" Artin's letter.[372]

Herzl. Grüße Ihre Emmy Noether.

Notes

[365] In the earlier letter from June 3, 1932 § 3.51. Noether had written that Hasse was listed for a lecture in the Mathematical Society of Göttingen. It seems that Hasse didn't know, and had asked Noether about it. Now Noether reports what happened.

[366] Noether means the paper[BHN32] on the Local-Global-Principle and the cyclicity of algebras.

[367] This seems to be a dig at Courant who on several occasions had expressed doubts whether the "modern algebra" of Emmy Noether is of true relevance for the future of number theory; on the other hand he was considered to have a wide overview over the mathematical sciences. We are reminded of the fact

that Courant in the early 1920s had advised Hasse against studying Hensel's
p-adics, because in his view this was an unimportant side track of number
theory. Hasse mentions this in the preface to his "Collected papers" [Has75];
it seems probable that Hasse had told this story to Noether.

[368] As already mentioned in the earlier letter of June 3, 1932 § 3.51, Hasse lectured only later, on January 13, 1933, at the Mathematical Society in Göttingen. But not about hypercomplex reciprocity laws, rather, corresponding to
his changed interest, on the Riemann Conjecture in function fields. We do
not know whether Hasse accepted Noether's offer to stay in her guest room.

[369] Noether refers to her previous postcard from June 14, 1932. § 3.54. There
she had requested that a reference to the notes of her lecture course 1929/30
be included. Now she clarifies that she would like the reference to be included
in Chevalley's paper and not necessarily in a paper of Hasse's.

[370] Hasse in his paper [Has33b] cites the contributions of Emmy Noether in
the following way: "E. Noether: *Lectures, in particular the lecture notes of
winter semester 1929/30, seminars, many discussions with her friends, in
particular with the author, letters to him 1929–1932.*"

[371] N.R.S. = norm residue symbol.

[372] See Noether's letter from April 5, 1932. § 3.47.

3.56   Postcard of July 21, 1932

*Ideal relations between a maximal order of a simple algebra and
the principal order of a maximal subfield.*

July 21, 1932

Sehr geehrter Herr Hasse!

Indeed I do not know your proof for the ideal relations[373] and
you do not know mine (which I first thought about when I heard
about yours in Bad Elster;[374] but as a simple application of structure theorems which I have had for a long time.) I suggest that
we publish both proofs, therefore three altogether, each in a short

note, in alphabetical order: Chevalley, Hasse, Noether. Chevalley had a bit more; and your proof was earlier than mine.[375] At first I was only interested in the structure of the maximal order with split crossed products, which is often convenient to know. Actually I wanted to use that for the Herbrand volume, but can just as well use, for example, ideal differentiation. So in the next few days I will send you this small note and get yours in return. Originally I wanted to prove the Principal Genus Theorem with these methods; I still consider this not impossible; for it seems to me that the Principal Genus Theorem is much more elementary than the now basic theorem on split algebras, in so far as it is essentially a matter of ideal classes, and not of divisors or elements.

You have noted and corrected Grunwald 107 instead of 106.[376]

Beste Grüße, Ihre Emmy Noether.

Notes

[373]This concerns the relations between the ideals of a maximal order of a central skewfield $D|K$ (where K is a number field or a local field) with the ideals of the maximal order of a maximal commutative subfield $L \subset D$.

[374]In September 1931 the DMV had held its annual meeting in Bad Elster.

[375]Of the three notes under discussion, those of Hasse [Has34d] and of Noether [Noe34], appeared in the issue of *Actualités scient. et industr.* dedicated to the memory of Jacques Herbrand while Chevalley published his note [Che34] in the *Hamburger Abhandlungen*.

[376]This refers to the corrections for the issue of Annals[Has33b] dedicated to Emmy Noether on her 50th birthday. Since Noether served as unofficial Annals editor, she apparently received the page proofs. Now she inquires whether one of the typos she found would be corrected. Grunwald's dissertation appeared in volume 107 of *Mathematische Annalen* [Gru32] and is now correctly cited by Hasse. Previously it read 106. – Wilhelm Grunwald was a doctoral student of Hasse. His name became widely known from the "Theorem of Grunwald-Wang", which plays a role in class field theory and the theory of algebras. Wang [Wan48] discovered an error in Grunwald's statement [Gru33] which Hasse had cited; this was corrected in [Wan50] and [Has50b]. See the the letter of November 22, 1931. § 3.39.

3.57   Postcard of August 03, 1932

Support for students. More about relations between commutative and non-commutative algebra.

August 3, 1932

Lieber Herr Hasse!

Today I spoke with Neugebauer about an assistant position;[377] unfortunately there is no hope that one will become free soon, since the budget for graduate teaching assistants had been frozen.[378]   On the other hand, Deuring told me that Van der Waerden has a student of Koebe in Leipzig in mind for his replacement, but there hasn't yet been any contractual agreement.[379]   But the replacement – one year – would only receive half of Deuring's salary, just over 100 M. Van der Waerden's address at the moment is Graz (Austria), Peinlichgasse 12, under [the name] Rellich.[380] — In the course of time there will be also the possibility of a stipend from the *"Notgemeinschaft"*.[381] I have obtained one [stipend] this summer for Fitting for 10 months.[382]

You will get my note next week; I am just now slowly working it out, and moreover in connection with your note, I have for the first time considered that I can also remove the assumption "k galois"; I still have to work that in.[383] What I have more than you[384] is that I can determine *all* ideals prime to the different, i.e., your $\mathfrak{p}$-components, by virtue of the module of all orders in k. At this time, I don't know whether it has number-theoretical significance; on the other hand, I am very interested in what you write about the connection with the Kronecker class number relation.

We have exchanged roles: you produce existence proofs, I write down explicit [formulas]; but all sorts of things will truly be in your transformed basis.

I would like to get [a copy of] Chevalley's proof; I don't have it.[385]

Along with my proof, [I am] returning your manuscr[ipt].

Herzl[iche] Grüße, Ihre Emmy Noether

Notes

[377]Neugebauer was acting chair of the Göttingen Math Institute in the absence of Courant. As such he was in charge of assistant positions. Probably Hasse had asked Noether whether there was an assistant position available in Göttingen. We don't know which of his students Hasse had in mind to send to Göttingen. There is a certain probability that it was Hans Reichardt who had just (1932) completed his doctorate with Hasse with a dissertation about cubic number fields [Rei33]. A few months later, in March 1933, Hasse asked Brandt in Halle whether he could accommodate Reichardt (which was not possible either).

[378]Note the date of this postcard. In those years there was a world wide economic crisis; German universities were confronted with budget shortages.

[379]At that time Deuring had an assistentship in Leipzig with Van der Waerden. He had been invited to spend one year with Oystein Ore at Yale University in the USA. This would temporarily make Deuring's position as assistant available. But we learn from Noether's letter that Van der Waerden had already made other plans for the replacement.

[380]"Rellich" was the maiden name of Mrs. Van der Waerden.

[381]The *"Notgemeinschaft für die Deutsche Wissenschaft"* was an agency of the German state, founded in 1920 in order to support research and in particular young researchers during those difficult economic times after world war I. In 1929 it had been renamed *"Forschungsgemeinschaft"* but Noether continued to use the old name.

[382]Hans Fitting had earned a doctorate with Noether in 1931 with a dissertation on automorphism rings. See the letter from February 8, 1931. § 3.25.

[383]This concerns Noether's paper "Split crossed products and their maximal orders" [Noe34] that was already mentioned in the previous postcard of July 21, 1932. 3.56

[384]Noether is referring to Hasse's paper "On certain ideals in a simple algebra" [Has34d].

[385]Probably she wishes to get the manuscript of Chevalley's paper [Che34] which was to appear in the *Hamburger Abhandlungen*. For, Hasse refers to Chevalley several times in his paper [Has34d], so Noether assumed that he had a copy of Chevalley's manuscript.

3.58   Letter of August 09, 1932

Noether sends her note, which appears later in the Herbrand-memorial volume. Rusam. Hopf.

Göttingen, August 9, 1932

Lieber Herr Hasse!

Here is my note, together with the return of other things. Since Chevalley is published separately – I had understood that he had sent you the manuscript for Crelle – we can also submit the notes to the Annals which at present publishes very fast and where I have not published for a long time. But Crelle too is fine with me if you prefer that; I leave it up to you. [386]

Are you making progress with your beautiful class number plans? The paper of Wenkoff ([translated] from the Russian in Math. Zeitschrift; I believe that is his name) about the determination of the class number of quadratic fields through the embedding in quaternion fields – represented through ternary forms – should be compared.[387]

I also enclose a letter from Krull's student Rusam – I don't need it back – which will interest you. For he has in the case of finite rings essentially solved the related Fitting problem. I will send him the Fitting manuscript that perhaps will be useful for him for a couple of general considerations; otherwise there isn't much. Possibly you can obtain further details from him.[388]

I think that Chevalley's enumeration at the ramified places can also be replaced by my explicit specification of the max[imal] orders, giving one more precise insight either proceeding as Chevalley by extending with splitting fields returning [us] to matrix rings, or by looking at the residue class ring modulo 𝔭, the two-sided prime ideal, which also is a matrix ring.[389]

I will probably leave at noon on Friday the 12th – the morning mail will still reach me – first to Wengen (Wengen, Berner Oberland, general delivery); perhaps afterwards I will go to Wengern-Alp.[390]

Deuring left already on Saturday for South Tirol, making a mountain tour with a friend (Meran, general delivery), from there to Zürich. What are your plans? Courant is expected at the end of August. In Zürich I will stay with Hopf, Schlößlistr. 2.[391]

Auf Wiedersehen dort und herzliche Grüße,

Ihre Emmy Noether.

Notes

[386]In her former letter of July 21, 1932 Noether had observed that Hasse, she herself and Chevalley had written notes about the same subject (it concerns the relation between the ideal theory of a a central simple algebra with the ideal theory of a maximal commutative subfield, the base field being a number field, global or local). She had proposed publishing these three notes together in some journal. But Chevalley had already submitted his note to the *Hamburger Abhandlungen*, and therefore she now proposes that at least Hasse's note and her own note should be published together, either in the Annals or in Crelle's Journal. Finally, however, both notes [Noe34], [Has34d] were eventually published in the Herbrand memorial issue of *Actualités scient. et industr.*. See also the postcard from July 21, 1932. § 3.56, as well as the following letter of October 29, 1932. § 3.59.

[387]We have been unable to determine which of Hasse's "class number plans" Noether had in mind. – B. A. Venkov investigated the arithmetic of quaternion algebras in the style of Hurwitz in the early 1920's. He came across an arithmetic proof of the Dirichlet Class Number formula for complex quadratic number fields with discriminant $d \not\equiv 1 \mod 8$ which he published in Russian in 1928 and in German in 1931 in the *Mathematische Zeitschrift* [Wen31].

His proof is also found in his book on number theory that appeared in English in 1970 [Ven70]. The stated condition on d arises because $-d$ must be a sum of three squares; this gives the connection with the embedding of the quadratic number field in a quaternion algebra.

[388]The dissertation of Friedrich Rusam (1934?) in Erlangen had the title: "Matrix rings with coefficients in finite rings of whole numbers." Apparently this work was not published in any mathematics journal; however, it was reviewed by Köthe for *Zentralblatt*. The letter from Rusam to Emmy Noether was found in Hasse's legacy. It seems that there was no exchange of letters from Hasse to Rusam.

[389]Noether refers to the note [Che34] and compares it to her own note [Noe34].

[390]It seems that Noether will spend some time in the Swiss Alps before the International Mathematician Conference in Zürich.

[391]In his time at Göttingen, Heinz Hopf (1894–1971) belonged to the inner circle around Emmy Noether, together with Paul Alexandroff. Since 1931 Hopf held a professorship in the *Eidgenössische Technische Hochschule* in Zürich.

3.59   Letter of October 29, 1932

Hasse's Visit in England.   The Herbrand Volume.   Principal Genus Theorem.   Artin's Conductor.   Chevalley.   Deuring's Report.   Springer and Abelian.

Göttingen, October 29, 1932

Lieber Herr Hasse!

I am glad that you have enjoyed your England trip; and thanks a lot for the card![392] For the Herbrand memorial volume[393] I actually wanted to use "Ideal differentiation and differents", in remembrance of my last conversation with Herbrand, who thought about the connections between the various definitions of different, if it can be made simple enough, which is still unclear.[394] But we have also spoken a lot about hypercomplex things so that is just as personal, and I also had in mind *"Maximal orders in crossed products"* for the Herbrand volume. So it seems to me now more proper to use that; it would be a shame to tear apart the connection between our notes. In this sense, therefore, in the next few days I will write to Chevalley, to whom I have not yet replied, except orally in Zürich. Perhaps you can submit our notes soon, since the things should be printed separately; I will ask Chevalley whether this makes sense.

Today I enclose two supplementary pages for the note, and ask you to incorporate them: the last line (formula line: $\mathfrak{G} = \mathfrak{H}S_1 + \cdots + \mathfrak{H}S_n$) on page 3 and page 4 until the end of the section; and the same for Note 6. The old page 4 must then become 4b. I have made the transition to the crossed product in the non-galois case somewhat more readable, and in particular corrected a few inaccuracies: the old u_S was defined incorrectly, and the explicit expression for the crossed product with arbitrary factor systems was wrong. Now I have not given it at all: it is called $K = \sum k v_S k$; if $\mathfrak{G} = \sum \mathfrak{H}S\mathfrak{H}$ and [we] set $v_S = \xi^{-1}\overline{u}_S\xi$. That probably holds analogously in characteristic p dividing k [395] by modifying your methods of proof in the case of k galois (going to subfields and extension fields and behavior of the factor systems); that should be done by a doctoral student[396] as well as how explicit factor systems [are] expressed and similar things.

I also enclose the "Principal Genus Theorem" intended for the Annals, with the request that a carbon copy be returned. If you have comments, please write these soon; I will wait several days before sending the manuscript to Blumenthal.[397] I have left out the conductor paragraph; it became so complicated that it didn't make sense anymore, since it wasn't used for anything, and I wasn't clear myself whether the ramified places came out by abelian specialization to high powers. I believe that progress will be made if one directly follows the relations between galois modules and crossed representations at every place; the conductor definition and relations to the Artin L-Series will arise by themselves by means of Deuring's theorems.[398]

Will the corrections to Chevalley's local class field theory come soon, which also belongs to this circle of questions? (Address of Deuring, who gets the corrections: Department of Mathematics, Yale University, New Haven (Conn.), USA).[399]

Now there is still a tragi-comic story to tell: Springer, and to be sure Ferdinand Springer, his imperial majesty, has declared that in the Annals galois and abelian will be written with capital initial letters; a paper of Ulm that has already been typeset must be changed. Yours will remain. For nearly a year I have in vain delivered to Blumenthal material; for example the uniform standard used by the number theorists, and platonic love! Even metabelian! If this matter sufficiently annoys you, you can complain to me so I can pass it on; success doubtful.

Nun herzliche Grüße, Ihre Emmy Noether.

Notes

[392]In October, 1932, Hasse visited his friend Davenport in Cambridge, and also Mordell in Manchester.

[393]After the sudden death of Jacques Herbrand, his friend Chevalley and André Weil planned to publish a book devoted to his memory. They had asked many mathematicians who had known Herbrand, including Noether and Hasse, to send contributions for this memorial volume. Hasse had suggested to Noether that they both send their notes on maximal orders to Chevalley for the Herbrand memorial volume, and Noether is now responding to that. She would have preferred to use another paper for the Herbrand memorial volume, but since that is not yet completed, she agrees with Hasse's proposal. It was planned that the book appeared in France in the publishing house Hermann. However, for reasons unknown to us, it turned out that all those contributions were not published in a single book, but they appeared each as a separate publication in the series "*Actualités scientifiques et industrielles*" of Hermann. For Noether's and Hasse's contributions see [Noe34], [Has34d]. See also the previous postcards. Compare also [Roq14].

[394]Noether never completed this paper "Ideal differentiation and differents". She had already announced it in 1929 at the DMV meeting in Prag. Her unfinished note appeared posthumously in 1950 in Crelle's journal [Noe50].

[395]These "supplementary pages" are meant for the manuscript of Noether's note [Noe34] which she had sent to Hasse who should forward it to Chevalley for publication (see above). In the published paper we read that k is a finite galois extension of a base field (which she calls Ω), and K is a split crossed product of the Galois group of $k|\Omega$ (which she calls $\mathfrak{G}$) with k. Although the main part of her paper is concerned with algebraic number fields and their completions, in the first paragraph of the paper she considers arbitrary base fields which may be of characteristic $p > 0$. In a footnote she stresses the fact that the characteristic p may well divide the group order, i.e., the degree of k over the base field. From this we conclude that this is meant when she writes to Hasse that "the characteristic divides k." – Although Noether writes that she considers also the case of non-galois extension, this does not appear in the published version; probably she had finally decided to leave this to her "doctoral student".

[396]Probably the doctoral student was Werner Vorbeck. Also see the letter from June 3, 1932. § 3.51.

[397]See [Noe33a].

[398]Noether had tried to define and investigate the Artin conductor in terms of characters of the galois group, with the help of hypercomplex systems. In the tamely ramified case, she carried this program out in the paper "Normal basis in fields without higher ramification"; this appeared in the Hensel-celebratory issue of Crelle's Journal [Noe32b]. However, the proof that Noether's conductor was the same as Artin's conductor (in the tamely ramified case) was given only much later, by Fröhlich [Frö83]. See also the letter of August 22, 1931. § 3.29. Noether was not successful in handling the case of wild ramification.

[399]This refers to the paper [Che33a]. Deuring, who was spending time in Yale supported by a scholarship, was supposed to take that paper into account in his book on algebras [Deu35a]. In fact, Deuring cited Chevalley's paper in his book but without going into detail, Deuring wrote only: *"Extending the proof of the reciprocity law and the theory of the norm residue symbol from cyclic to abelian fields offers no difficulties"*. A particular easy proof was published by Witt in the year 1935 [Wit35b]. See also the letter of June 3, 1932. § 3.51.

3.60   Postcard of November 25, 1932

Hasse's Marburg Lecture Notes on Class Field Theory. Scholarship for Olga Taussky. Hasse's Critique of the Principal Genus Theorem Manuscript.

Göttingen, November 25, 1932

Lieber Herr Hasse! [400]

I have, after Miss Bannow[401] brought me a carbon copy, still three more orders of your Class Field Theory.

> Wichmann, Cabrowstr. 10
>
> Dr. Gröbner, Lotzestr. 24
>
> Knauf, Gronerstr. 14.

If you prefer, send all together to me! I hope that sufficiently many have signed up so that it won't be too expensive for the students, in any case not more expensive than the stated 7–8 M, otherwise I would have to request a price announcement in advance.[402]

Many thanks for your critique of Principal Genus Theorem. I have made small additions in various places and now hope that it is complete, since the hypercomplex is still nontrivial.[403] Whether galois or Galois, I don't know!! You should convince yourself from the page proofs!

I have omitted the conductor. That will come later together with other things, when it has been more thoroughly worked through. And I wanted to emphasize that anything goes *without* the conductor.

Has Miss Taussky written to you that with my advice she has applied for a scholarship from the "International University Women"? You are listed as a reference; she listed Germany as the country where she will work. But you can also write a letter of recommendation on your own volition to the German Association of Women Academics, care of Frau Dr. Schlüter-Hermkes, Berlin-Charlottenburg, Lietzenseeufer 8.[404]

Herzlichst, Ihre Emmy Noether.

Notes

[400]As usual, this postcard was addressed to Hasse's address in Marburg but was then sent to Fraenkel's address in Kiel where Hasse was staying during his visit. Hasse had been invited to a colloquium lecture in Kiel. This lecture was the first in which Hasse spoke on the number of solutions of diophantine congruences; this topic led him 4 months later to the proof of the Riemann Hypothesis for elliptic function fields with finite base field. See [Roq18].

[401]Erna Bannow, the later Mrs. Witt, was a mathematics student.

[402]In summer semester 1932 in Marburg Hasse had given a lecture course presenting the new state of class field theory. The lecture notes were prepared under the guidance of Wolfgang Franz, a Ph.D. student of Hasse. (See the letter of April 5, 1932.) The copies of the notes were going to be produced on a hectograph to make them available to a larger circle of interested people. To estimate the needed number, Hasse had sent out a newsletter requesting pre-orders. These notes became known as the "Marburg Lectures" [Has33d] and enjoyed great popularity world wide among number theorists. They appeared later in book form [Has67].

[403]Together with her letter of October 29, 1932, Noether had sent her manuscript on the Principal Genus Theorem [Noe33a] to Hasse with the request for his comments.

[404]In 1931/32 Olga Taussky had worked in Göttingen on the editorial staff of Hilbert's Collected Works [Hil32] [Hil33] [Hil35]. (See the letter from March 15, 1932.) Apparently her position had run out and Taussky returned to Vienna. Noether took it on herself to look for new financial support for her. The MacTutor History of Mathematics Archive maintained at St-Andrews University says this about Olga Taussky: *"Leaving Göttingen in the summer of 1932, she received a letter from Courant before the new academic year started advising her not to return to Göttingen due to unrest at the university caused by the deteriorating political situation."* Accordingly, she didn't accept the scholarship to study in Göttingen after it was approved.

Later, in 1981, Olga Taussky-Todd wrote a moving article *"My personal recollections of Emmy Noether."* [TT81].

3.61   Postcard of November 30, 1932

Hasse invited to lecture in Göttingen. Ackermann-Teubner Prize for Noether and Artin.

Göttingen, November 30, 1932

Lieber Herr Hasse!

On behalf of the Math[ematical] Society, I would like to ask whether it would suit you, in 14 days, Tuesday, December 13, to speak, or whether you would prefer the beginning of January. Yes, a lecture was made out for this winter![405] This time you must really stay with me. Aleksandroff thoroughly sanctified the guest room with a four week stay. You arrive as usual at 5^h03 p.m. in the afternoon? The Math Society, at Weyl's request, against all tradition, set it at $5 - 7^h$ p.m., so your lecture must be set for 6^h p.m. Or would it be alright to lecture right from the train station?

Artin and I, as Ackermann-Teubner (Algebra-Number Theory)-Prize winners, (with the entire 500 M each) are supposed to lecture in Leipzig around Christmas; I think simultaneously, but haven't yet understood this from Van der Waerden's letter.[406] But that will be at the end of the week, therefore does not collide with the Math[ematical] Society.

Don't you want to give a separate lecture for me on Wednesday afternoon? I have developed up to now a systematic theory of p-adic numbers – doesn't that impress you? – as foundation for non-commutative arithmetic, which in the following semester should devolve into commutative applications.[407]

Will the corrections of Chevalley's local Class Field Th[eory] come soon?[408] Also for Deuring who is now [working] really seriously on his Report; he feels very well in America.[409]

Auf Wiedersehen, Ihre Emmy Noether

Notes

[405]Hasse finally went to lecture in Göttingen in January, 1933.

[406]This refers to the "Alfred Ackermann-Teubner memorial prize for supporting the mathematical sciences". If Van der Waerden is mentioned in this connection, then one can conclude that he belonged to the nominating committee for this prize. Van der Waerden was at that time full professor in Leipzig, the seat of the Teubner publishing house.

[407]In winter semester 1932/33 Emmy Noether lectured on "Non-Commutative Arithmetic" , 2-hrs., Wednesday 5–7. For the following summer semester 1933 she announced "Hypercomplex Methods in Number Theory" – but due to political developments this didn't come to pass.

[408]Noether is referring to Chevalley's Crelle paper [Che33a].

[409]For Deuring see the notes to the letter of August 3, 1932. §3.57.

3.62   Postcard of December 09, 1932

Hasses lecture set for January 1933.   Zeta function in title to peak interest. Ackermann-Teubner Prize is reduced by half due to "economic hard times".

December 9, 1932

Lieber Herr Hasse!

Your lecture is set for Tuesday, January 10, 33, so as soon as possible after Christmas. I'll expect you then around 2^h. Send me the exact title of your talk, better already now *before* the holiday. You can mention the zeta function to peak interest. [410]

My lecture time is Wednesday $5 - 7^h$; naturally can be moved up if you want to leave in the evening; but I think you [should] stay at least to Thursday; that is more worthwhile. I have discussed the transition from local to global in great generality (i.e. your various conclusions [from] Ann[alen] 104 summarized in general module theorems); the people are therefore in fact well prepared for your lecture.[411]

I will lecture in Leipzig on Thursday the 15th; Artin unfortunately only at a later time, since Van der Waerden's travel funds are insufficient for both. Moreover the prize [money] amounts to only 250 each; due to the "current crisis" they have reduced us by half.[412]

I want especially to emphasize the viewpoint of the Principal Genus Theorem as a theorem on representations; I believe that in this direction one can come to reasonable formulations whereby the Brandt groupoid of ideal classes will give the formalism – analogon to the "multiplicative basis" Dc_{ik} of the matrix ring D_r, which likewise [is a] groupoid (c_{ik} matrix units).

Blumenthal has forwarded the "Hasse war declaration" to Springer; hopefully this will at least have achieved that your dedication paper "Abelian" will not be retroactively spoiled (as they have done in Fitting's case).[413]

Beste Grüße, Ihre Emmy Noether

Notes

[410]Hasse's new area of interest, to which he was introduced by his friend Davenport, is the estimation of the number of solutions of diophantine congruences. In November 1932, he had lectured on that in Kiel and in Hamburg.

Artin pointed out to him that the problem is closely related to the analogue of the Riemann hypothesis for the zeros of congruence zeta functions. Apparently Hasse had already written this to Noether, and she is of the opinion that putting 'zeta function' in the title would attract a larger audience. – By the way, only 2 months later Hasse found his first proof of the Riemann hypothesis for *elliptic* congruence zeta functions. For more about this, see [Roq18].

[411]On Wednesday Hasse spoke in the Noether seminar about the Local-Global Principle for quadratic forms.

[412]This refers to the Ackermann-Teubner prize, which Noether already mentioned in the postcard of November 30, 1932.

[413]As Noether reported in the letter of October 29, 1932, it was a question of whether the Springer journals should write "Abelian" or "Galois", and so forth, with an initial capital letter; at Noether's urging Hasse had written a protest letter; probably he demanded that the decision about such matters be left to the authors. We do not know how this "war" ended.

3.63 Postcard of December 11, 1932

Theme for the Göttingen lecture. Music of the future: Groupoid isomorphisms and inverse theorems for galois fields.

December 11, 1932

Lieber Herr Hasse!

Many thanks for the message with the lectures! [414]

Don't you want to come already on Monday the 9th, say with the 5pm train, since you have lectures in the morning. That would be very worthwhile! Or are you also scheduled up Monday afternoon? If you answer me right away I can announce your exact plans already on Tuesday (the 13th) in the Math[ematics] Soc[iety]. Possibly we can then also set the lecture [in my seminar] on Monday evening, 6 − 8pm: then we'd have Wednesday after dinner free for a walk. So let me know your plans!

In the meantime I have thought more about the groupoid matter. I'm inclined to the point of view that the inverse theorems for arbitrary galois fields will lead to such groupoid isomorphisms – or what is the same thing, mixed group isomorphisms, so operator isomorphisms.[415]   And to be sure then it must concern relations between the composition of the absolute irreducible representations of the group, and the crossed representations in the ideal class group.   Here it is allowed not only to compose, as in the abelian case, the (two-sided) representation classes (characters), [but] rather the representations themselves (so the one-sided classes) with the representation classes – in analogy with the composition of one-sided and two-sided ideal classes in simple algebras.

But that is music of the future, and perhaps not even true.

Beste Grüße, Ihre Emmy Noether

Notes

[414]The theme for the lecture for the Mathematical Society was *"On the zeros of the Artin congruence zeta function."* One day later Hasse spoke in the Noether seminar with the theme: *"On the theory of quadratic forms"*. It concerned the local–global principal for quadratic forms.

[415]Here Noether is thinking again about the generalization of class field theory to arbitrary galois (not necessarily abelian) extensions. The concept "groupoid" was introduced by Brandt in connection with the description of the ideal theory in maximal orders in simple algebras. It was further developed by Artin [Art28b], [Art28a]. We do not know what Noether had in mind when she suggested that this concept could be brought in connection with non-abelian class field theory.

3.64 Letter of December 26, 1932

Answering questions about hypercomplex systems.

Lieber Herr Hasse! [416]

I have now thought about some but not all of your questions. I begin with the last one:

Let $P = (S(e_i e_k))$ be the system matrix: then $e = \mathfrak{e}P^{-1}\mathfrak{e}'$ the principal unit, provided P is negative. It is known that P negative means that the system is semi-simple, and that the fundamental numbers of the absolute irreducible representations are not divisible by the characteristic of the coefficient domain. So in particular in character[istic] zero the second condition is satisfied. Now $\bar{\mathfrak{e}} = \mathfrak{e}P^{-1}$ becomes the complementary basis to $\mathfrak{e}$: for

$$\mathrm{Sp}(\mathfrak{e}'\bar{\mathfrak{e}}) = \mathrm{Sp}(\mathfrak{e}'\mathfrak{e}P^{-1}) = \mathrm{Sp}(\mathfrak{e}'\mathfrak{e})P^{-1} = PP^{-1} = E.$$

Complementary bases transform contragrediently: Let $\mathfrak{c}' = C\mathfrak{e}'$; so $\mathfrak{c} = \mathfrak{e}C'$: then $P_{\mathfrak{c}} = CPC'$; (Section 25 of my representation paper), so $\bar{\mathfrak{c}} = \mathfrak{c}P_{\mathfrak{c}}^{-1} = \mathfrak{e}C'^{-1}P^{-1}C^{-1} = \mathfrak{e}P^{-1}C^{-1} = \bar{\mathfrak{e}}C^{-1}$. With that $\bar{\mathfrak{c}}\mathfrak{c}' = \bar{\mathfrak{e}}\mathfrak{e}'$ follows. It suffices to verify that $E = \bar{\mathfrak{c}}\mathfrak{c}'$ for any basis.

I choose $\mathfrak{c}$ as the system of matrix units (after passing to a suitable extension field of the coefficient domain) $c_{ik}^{(\nu)}$. The corresponding complementary basis is: $\bar{\mathfrak{c}} \sim c_{ki}^{(\nu)} \cdot \frac{1}{n^{(\nu)}}$ ($n^{(\nu)}$ [is the] degree of the individual matrix algebras). Therefore

$$\bar{\mathfrak{c}}\mathfrak{c}' = \sum_{\nu}\sum_{i,k} \frac{1}{n^{(\nu)}} f_{ki}^{(\nu)} c_{ik}^{(\nu)} = \sum_{\nu}\sum_{k} c_{kk}^{(\nu)} = \sum_{\nu} e^{(\nu)} = 1.$$

The considerations about the contragredientness of the basis $\mathfrak{c}$ are apparently preserved if one bases [them] on the reduced trace matrix, in which case $c_{ik}^{(\nu)}$ and $c_{ki}^{(\nu)}$ are complementary. But the trivial representation only comes out if one row and column of $e_{ik}^{(\nu)}$ is taken.

The considerations about the complementary basis give another meaning to the second Frobenius relation: $PM_\alpha = N_\alpha P$, if again P is taken to be the trace matrix. For $PM_\alpha P^{-1} = (P^{-1})^{-1}M_\alpha(P^{-1})$ then gives exactly the representation which comes from the complementary basis $\bar{\mathfrak{e}}$, which is therefore equal (not only equivalent) to the right representation $\mathfrak{e}'a = N_\alpha\mathfrak{e}$.

Due to the contragredient transformation, one can restrict to bases $\mathfrak{c}'$, $\bar{\mathfrak{c}}$: but here it is clear that $c_{11}^{(\nu)},\ldots,c_{1n_\nu}^{(\nu)}$ produces the same representation as $c_{11}^{(\nu)},\ldots,c_{n_\nu,1}^{(\nu)}$. The relation holds therefore also with the reduced trace.

Finally, the first Frobenius relation: $N_{\beta'}M_\alpha = M_\alpha N_{\beta'}$ means that the system, viewed as a right module, is reciprocally isomorphic to its automorphism ring (full, when there is an identity). For, the representation of the (operator-) automorphism ring is given through the totality of the matrices that commute with the representation (this can be found in many places, e.g., in my journal article " On splitting fields" and so forth). The fact that the full automorphism ring is identical with the theorem of Frobenius is §1 and more generally in §3 (where the existence of the identity element is assumed).

Forming the direct product of the system $\mathfrak{S}$ with the aut. ring $\overline{\mathfrak{S}}$ (or reciprocal isomorphism) is also connected to the peristrophic matrix formed with "parameter" $\mathfrak{x} = \xi_1,\ldots,\xi_n$. For, if one looks at $\xi_1,\ldots,\xi_n$ as basis units of an isomorphic system, the per. matrix P_ξ transforms into $(\xi_i\xi_k) = \mathfrak{x}'\mathfrak{x}$. One can then define a generalized complement through $\bar{\mathfrak{e}}_\mathfrak{x} = \mathfrak{e}P_\xi^{-1}$, if P_ξ^{-1} is defined in $\mathfrak{S}\times\overline{\mathfrak{S}}$, where $\overline{\mathfrak{S}}$ coefficient domain.

Again, complementary bases transform contragrediently; and now [we only need] the weaker assumption of the existence of an inverse to the matrix formed with parameters (in $\mathfrak{S}$); it seems that the two-sided non-split components of $\mathfrak{S}$ become primary so the splitting of the residueclass rings correspond to radicals. One still has to be able to make the further conclusions with the complementary basis: I did not succeed right away with the first consideration. It must have to do with an analog of my multiplication

Z_Z of a commutative galois field with itself, and around the connection of idempotents and complementary basis in the ordinary sense, as it appears in Z_Z. Perhaps you have someone who can translate Frobenius into this language; it definitely is worthwhile!

I wish you the necessary snow for skiing; it doesn't look promising. I want a few days (probably Thursday through Tuesday) in Berlin and Magdeburg to friends and family.

A good 1933! Auf Wiedersehen in January,

Ihre Emmy Noether.

Notes

[416]This is Noether's reply to a letter by Hasse in which he asked questions about some details of a manuscript of Noether. We do not know those questions and were not able to find Noether's manuscript. It may be that the manuscript was an early version of her paper which appeared in the Herbrand memorial volume [Noe34] but we are not sure. So we decided to leave this letter uncommented.

3.65   Postcard of February 2, 1933

Scheduling Noether's colloquium lecture in Marburg. Comments on Zorn's proof. Corrections for Hasse's work on skew fields.

February 4, 1933

Lieber Herr Hasse,

First of all, I wish that you will soon be able to dismantle your family-hospital! For me, February 17 is naturally just as good –

I will arrive in the afternoon.[417] In the meantime I have finished some things; it still isn't a lot.

Zorn's proof for split algebras brought me much pleasure; to me that looks more and more like a hypercomplex Riemann-Roch theorem, one should be able to prove this without *zeta*-functions! [418]

I bumped into a few trivial errors while reading your paper on skew fields; I will give them to you in Marburg. Instead of referring to the previous theorems, you write down explicit splittings, where the ideal factors are not integral.[419]

A good recuperation to everyone!

Herzliche Grüße, Ihre Emmy Noether.

Notes

[417]Noether had been invited to visit Marburg for a colloquium talk, but due to some health problems in Hasse's family (perhaps a common cold) the date of the visit had to be shifted.

[418]Zorn, a former student of Artin, showed in a paper in the *Hamburger Abhandlungen* [Zor33] that the local-global principle for algebras over number fields can also be derived from the dissertation of Käte Hey. Although Zorn's paper had not yet appeared, Noether seems to know the manuscript. Käte Hey earned her doctorate in 1927 with Artin in Hamburg. Her dissertation deals with the zeta function of skewfields over number fields [Hey29]. The paper was never published in a journal, but was printed and passed around to interested mathematicians. Hasse and Noether had received a copy. Hey's paper contained errors, was difficult to read, and the zeta function needed to be completed with factors at the infinite places. (See also [Lor05].) That is what Zorn did in his paper. When Noether speaks of the "Riemann-Roch Theorem" she is referring to the dissertation of Witt 1933. Witt had been "deeply impressed" (as he said himself) with Artin's Göttingen lectures on class field theory in February 1932, and this stimulated him to develop class field theory for function fields with finite constant fields. Noether gave him the dissertation topic of carrying Käte Hey's methods and results over to

function fields. He did this in his dissertation [Wit34a], whereby he formulated and proved the "Riemann-Roch Theorem for algebras" in the function field case. This theorem has become significant in connection with so-called non-commutative algebraic geometry. – It is interesting to note that Noether immediately had the idea that one should be able to formulate the "Riemann-Roch Theorem for algebras" also for number fields.

[419]This concerns Hasse's paper on p-adic skew fields [Has31d].

3.66   Letter of March 3, 1933

Congratulations on "Riemann Conjecture". Note from Chevalley on arithmetic on matrix algebras. Hasse's work on explicit construktion of class fields.

March 3, 1933

Lieber Herr Hasse!

Above all, my congratulations on "Riemann conjecture". You have accomplished unbelievably much recently! I conjecture that you will now approach the general Artin-Schmidt zeta function since you already are attracted to general class field theory.[420]

The note from Chevalley interests me greatly; I conjecture that such an appearance of ideals as coefficient domain ($\mathfrak{a}_{ik}$ coefficient domain of e_{ik}, the unit matrix) will play a role everywhere with application of crossed products, with the characterization of the maximal order, that I only have with factorsystem One, in my Herbrand note. Perhaps one will come to further meaning of the ideal factorsystem.[421]

Your class field theory has now become clear in questions and methods.[422] I am still stuck on a place, that is due *only* to my clumsiness. You conclude that α_0 is a principal ideal apparently from the

Lemma: If the principal ideal $(\alpha) \underset{n}{=} \mathfrak{p}_1^{\bar{a}_1} \cdots \mathfrak{p}_s^{\bar{a}_s}$, then there is a principal ideal $(\alpha_0) = \mathfrak{p}_1^{a_1} \cdots \mathfrak{p}_s^{a_s}$ with $a_i \equiv \bar{a}_i \pmod{n}$.

That must follow from simple conclusions about absolute ideal classes and abelian groups; but I don't see it. Or does one somewhere find this conclusion from the n-equality to actual equality? I have the impression that I am not seeing a triviality.[423]

Herzliche Grüße, Ihre Emmy Noether.

Notes

[420]This concerns the proof of the analog of the Riemann conjecture for elliptic function fields with finite constant field. From Hasse's correspondence with Davenport and Mordell we know that Hasse found the proof at the end of February 1933. Soon after he informed Emmy Noether. Hasse's proof used the class field theory of complex multiplication which he had described in earlier articles in Crelle's journal: [Has26d], [Has31b]. That is why Emmy Noether mentioned class field theory. When Noether speaks of the general Artin-Schmidt zeta function, she is referring, first, to Artin's dissertation (1921) in which he defined the zeta function for hyperelliptic function fields, and second, F. K. Schmidt's habilitation paper (1927) in which the zeta function of arbitrary algebraid function fields (with finite constant field) were defined and investigated. See [Art24a], [Art24b], [Sch31a].

[421]This concerns Chevalley's paper entitled "Sur certains idéaux d'une algèbre simple" [Che34], which appeared in the Hamburger Abhandlungen. Perhaps Noether (and Hasse) had a copy or a preprint or the galley proofs. There are close connections of this paper with the papers of Noether [Noe34] and Hasse[Has34d], which appeared later in the memorial volume for Herbrand.

[422]In this note, Hasse gives an explicit construction of cyclic class fields under the assumption that the relevant roots of unity lie in the ground field. On February 22, 1933, Hasse sent the manuscript to Emmy Noether (in her capacity as (unofficial) editor of *Mathematische Annalen*). The paper [Has33c] appeared in 1933. When Noether writes that the paper has "now"

become clear, we can assume that there was an eaarlier version that was not so clear. However, no earlier version is known to us.

[423]That was a printing error in the manuscript. See the following letter from March 22, 1933.

3.67 Letter of March 22, 1933

Noether's trip to Westerland. Correction of Hasse's class field theory manuscript. Commentary on Noether's talk in Marburg. Deuring: Zeta functions of quadratic forms. The role of transfer in the principal ideal problem.

March 22, 1933

Lieber Herr Hasse,

Since on Sunday I will travel for a few weeks (Westerland – Sylt, Villa Richard), I would like to ask [you] to send the lecture notes (also for Tsen) only at the beginning of the semester.[424]

By the way, in both versions of your class field theory manuscript you really had the incorrect formula: $\prod_{\mathfrak{p}|\mathfrak{f}} \mathfrak{p}^a = \alpha_0$ (principal ideal); that's why I didn't understand! I have removed the $\mathfrak{f}$ and changed the text a bit; of course you can make what changes you like in the page proofs![425]

In Marburg it seems that I claimed more than I can prove; the assignments are not so smooth and clear, such as the representation of the icosahedral group would make you think.[426] At the moment, it does not proceed.

Deuring, for the Herbrand memorial volume, in continuation of his analytic investigations, has given the following theorem: "There is a positive number a so that the function $Z_d(s) = \sum(n^2 + bnm + fm^2)^{-s}$, where the form has discriminant $-4d = b^2 -$

$4f$, in the domain $0 < t < d^a$ has only zeros with real part $\frac{1}{2}$."
That seems yet again to be a very pretty thing! He will speak
about that soon in New York at the meeting of the Amer. Math.
Soc.[427]

I have thought about the following interpretation of the "transfer"
in the Principal Ideal Theorem. If one forms the crossed product
of the ideal cl[ass] group (commutator gr[oup]) with the corresponding abelian (galois) group, and then [takes] such a power so
that the factor systems to $S_2, \ldots, S_n$ equal 1, then [the factorsys-
tem] to S_1 becomes the transfer of S_1. According to the theorem
of Chevalley and you. That must mean a reduction to cyclics![428]

Beste Grüße, Ihre E.N.

Notes

[424]Observe the date of the postcard. Since the past two months Germany
had a Nazi government. There were antisemitic riots in Göttingen. For
the time being, the university was closed; the semester start, traditionally
shortly after Easter, was postponed until May, 1933. This explains why
Noether was able at the end of March to leave Göttingen "for a few weeks"
and travel to her vacation resort on the North Sea. – The lecture notes that
she mentioned are Hasse's "Marburg Lectures" on class field theory[Has33d]
from summer semester 1932. See the postcard of November 25, 1932.?? –
Chiungtze C. Tsen was a doctoral student of Emmy Noether. In his doctoral
dissertation he found the highly regarded "Theorem of Tsen", according to
which every algebra over an algebraic function field of one variable splits if
the field of constants is algebraically closed.

[425]See the previous letter from March 3, 1933.3.66 The error did not appear
in the printed version.

[426]We do not know what Noether spoke about in Marburg. Since she mentions the icosahedral group, it seems not unlikely that it dealt with the theme
"class field theory for non-abelian galois extensions." As seen in earlier letters, Noether worked for more than a year with this theme, encouraged by
Hasse's result, which was mentioned in this connection in the joint paper of
Brauer-Hasse-Noether [BHN32]

[427]Deuring's results concern zeta functions of quadratic forms. Noether's description of his main result uses other, unexplained notations than Deuring himself, and hence it is not easy to understand. For a precise description of Deuring's results we refer to his paper [Deu35c] which appeared in Crelle's Journal (and not in the Herbrand memorial volume as Noether had announced). A preview was published in the Bulletin of the AMS 1933.

[428]The "Theorem of Chevalley and Hasse" that is mentioned here is the one that Noether discussed in her letter of June 3, 1932. It seems that Noether tried to use that theorem for a proof of the Principal Ideal Theorem of class field theory, by reduction to the case where the commutator factor group is cyclic. But as she said in her letter, her idea did not work.

3.68 Letter of May 10, 1933

Noether put on "leave". Plans for Princeton. Noethergemeinschaft. Schilling. Chevalley. Principal Genus Theorem.

May 10, 1933

Lieber Herr Hasse,

Many many thanks for your good empathetic letter![429] The situation for me is much less terrible than for very many others: I still have small assets (I never was eligible for a pension), so I can wait in peace at the moment until a definite decision [is made], or a bit longer, [if] the salary goes further. Some of the faculty have tried to make the dismissals temporary; the success is naturally quite questionable at the moment.

Moreover Weyl said to me that already a few weeks ago when everything was in flux, that he had written to Princeton where he still has connections. To be sure, due to the dollar crisis they now also have no decision power; but Weyl thinks that in time something could come of this.

Because of that, last year Veblen put a lot of importance in introducing me to Flexner, the organizer of the new institute. [430]

Perhaps one day a possibly repeating guest lecture will come from that, and moreover Germany again, that naturally would be best. And perhaps someday I and also you can arrange a year at the Flexner-Institute – that is future-fantasy to be sure – but we spoke of that in winter.

I thought about writing to Ore and Deuring if Richard Brauer needs something, to be sure he is still in a position as assistant, but he isn't allowed to teach "at the urgent recommendation" of the culture minister "for the time being"; Ore had already thought about inviting R. Brauer for a year, but then the needed funds were not available.[431] This "not teaching for the time being" is rather catastrophic here in the Institute; perhaps Davenport has already informed you.[432] But for all of that certainly rather quickly a calming will come![433]

I am reading your draft with great pleasure;[434] I think that from time to time I will gather the *"Noethergemeinschaft"* in my living room to discuss it.[435] Schilling[436] wrote to me about Chevalley's new approaches to hypercomplexes; it looks like what I always wanted, and I am very excited about it!

Herzliche Grüße, Ihre Emmy Noether.

At the same time, I am sending to the seminar the Principal Ge[nus] Theorem.[437]. Your critique is fully taken into account.

Notes

[429]On April 25, 1933, Emmy Noether was put "on leave effective immediately" by a telegram from the [culture] ministry, along with five other Göttinger scientists. Thus Emmy Noether was one of the first to be hit by the full power of the antisemitic Nazi law which was aimed at eliminating people of Jewish descent from public service. Against this background, since Noether thanks Hasse for his "good empathetic letter", we can assume that Hasse wrote her a letter expressing his continuing friendship, when he heard

about her enforced "leave". We do not know details of Hasse's letter. In particular, we don't know whether Hasse already in this letter discussed his plan to collect expert opinions in an attempt to get her dismissal reversed. That gets discussed in later of Noether's letters.

[430]Abraham Flexner was the spiritual father and first director of the Institute for Advanced Study in Princeton, that Noether calls the "Flexner-Institute". – It is likely that Emmy Noether got together with Veblen at the International Mathematicians Congress in Zürich in September 1932. Flexner was also participating at this Congress, and Noether met him there.

[431]Øystein Ore at Yale University had close scientific contact with Emmy Noether for many years. Together they edited the collected works of Dedekind [Ded32] (with the participation of Robert Fricke, who died before the book was published). Eventually it turned out that R. Brauer went to Princeton where he got a position as assistant to Hermann Weyl. – Max Deuring, who Noether had called "her best student", remained on scholarship at Yale with Ore.

[432]In the summer of 1931 Davenport had lived with Hasse's family, and there developed a close friendship. For the summer semester 1933 Davenport received a stipend from his college in Cambridge (Trinity) for a research visit in Göttingen. Since the semester start had been pushed to May 1933, in April Davenport lived with Hasse in Marburg, where they worked together on the proof of the Riemann Conjecture for the curves which today are called "Davenport-Hasse curves". In May 1933 Davenport moved to Göttingen and was witness to the dissolution of the Mathematics Institute as a consequence of Nazi politics. On weekends Davenport used to travel to Marburg, hence Noether assumed that Hasse had already heard details of the conditions in Göttingen from Davenport.

[433]As did many others at that time, Noether thought that after a certain time normal conditions would return. She didn't foresee the catastrophic developments of the next twelve years.

[434]This refers to the notes of Hasse's Marburg Lectures from the year 1932 [Has33d]; see Noether's previous letter from March 22, 1933.3.67

[435]Noether was no longer allowed to hold her seminar in Universitiy buildings. As we learn, she intended to gather her inner circle in her apartment to work through Hasse's draft together. And she eventually did so. It is told that her student Ernst Witt once attended this gathering in an SA-uniform, but that did not disturb Emmy Noether. (This is cited in Kersten's biography of Witt [Ker00]; it is also mentioned by Hermann Weyl in his speech at her funeral, who said that "she never for a second" doubted his integrity. See [Roq10], Chapter 3.2.) – The name "*Noethergemeinschaft*" was chosen as a take-off from the name "*Notgemeinschaft*"; see the postcard of August 3, 1932. §3.57.

[436]O.F.G. Schilling was a doctoral student of Noether's. After Noether had to emigrate, Hasse took Schilling as his student. In 1935 Schilling received a doctoral degree in Marburg with a dissertation on the theory of algebras. Later Schilling emigrated to Cambridge (England) and then to Princeton (USA) at the Institute for Advanced Study, on Hasse's recommendation to Weyl.

[437]This concerns the page proofs of Noether's paper on the Principal Genus Theorem [Noe33a]. Noether had asked Hasse for his review and critique. See the letter from November 25, 1932. § 3.60

3.69   Letter of June 21, 1933

Report on Noether. Hypercomplex interpretation of class field theory. Ideal factor systems. Elementary proof of Hasse's Existence Theorem. Chevalley's thesis. Tsen's Theorem.

June 21, 1933

Lieber Hasse! [438]

Many thanks for your letter from *Pfingsten*;[439] you really took on a big job with the expert reports! [440] And you already had enough work! Wichmann had given the student signatures[441] – it was mainly the algebraists – just in time to the curator who traveled to Berlin this *Pfingsten*, who was in agreement with it although naturally at the moment it would be difficult to get over § 3.[442] And yes he got your letter too! [443] It seems to me that it would be good, if enough recommendations arrive by the end of the semester, to send them to the curator[444] and to send late arrivals (Takagi, etc.) afterwards; word is that no decision will be taken earlier in the semester.[445] And it also seems right to me to copy the previously written reports (at my expense!) in order to be able to possibly return to them later, in case you don't have success at the moment.

Sending the version "Commissioned with special lectures now or later" together with the recommendations seems to me the best way.[446]

And now I hope not to come with new wishes.

In the hypercomplex interpretation of class field theory[447] I can now also capture the ideal factorsystems – therefore all classes of A/H in the cyclic case. I form the module $\mathcal{O} + u\mathcal{O} \ldots + u^{n-1}\mathcal{O}$ with $u^n\mathcal{O} = \mathfrak{a}\mathcal{O}$; then (and only then) can this be embedded in an algebra. That is, the rules of calculation in the module, and in particular $u^n\mathcal{O} = \mathfrak{a}\mathcal{O}$, are induced via the calculation rules of $Z + uZ + \ldots + u^{n-1}Z$ with $u^n = \alpha$, where the algebra splits at the ramified places of Z, if $\mathfrak{a}$ lies in H. (If $\mathfrak{a} = N(\mathfrak{A}) \cdot \alpha$, then $v = u\mathfrak{A}$ gives the module $\mathcal{O} + v\mathfrak{A} + v\mathfrak{A}\mathfrak{A}^S + \ldots$, which lies in $v^n = \alpha$, and, the other way around, this gives all $\mathcal{O}$-modules such that v is associated with u in the crossed product of the ideal group $\mathfrak{J}$ with $\mathfrak{G}$, therefore in the extension of $\mathfrak{J}$ with $\mathfrak{G}$. Conversely every differential system can be generated in this way using a suitable cyclotomic field. (Existence Theorem (0.3), Ann. 107). The Reciprocity Law says only that a "proper integral" over Z remains proper over a generating cyclotomic field; i.e., that the corresponding differential system is generated by a cyclotomic-field-integral and not by an algebra. Perhaps that can be proved directly using the method of "crossing" [with a cyclotomic field]! And then [we get] everything without the Norm Theorem. I have not yet thought [deeply], can not yet vouch for the correctness of this sketch!

In this connection, I am interested [in knowing] whether your Existence Theorem (0.3) has an elementary proof in general, as you did for for $\nu = 1$ in your manuscript.[448]

Furthermore, I would like the corrections or a carbon copy of Chevalley's thesis, if one exists; for I have the impression that the sketch above is only his considerations dressed in other clothing.[449]

Did you know that over algebraic function fields in one variable (perhaps also for n variables) with algebraically closed constant field, in particular with complex coefficients, no [non-trivial] division algebra exists; so the theorem of split algebras holds trivially? The Chinese [student] Tsen has now proved it following my suggestion.[450] One simply shows by comparing coefficients and equation counting that every element of the ground field is a norm; then one makes our transfer from cyclic to general.[451]

Herzliche Grüße, Ihre Emmy Noether

Notes

[438]Note the changed greeting from earlier and later letters, where it always was "Lieber Herr Hasse". According to the custom of the time, the omission of the formal "Herr" was evidence of a special empathetic friendship.

[439]Literally this refers to Pentecost, but it is used here in a non-religious sense to indicate an annual public holiday occurring around seven weeks after Easter.

[440]Hasse was in the process of gathering affidavits from well-known colleagues, also from abroad, attesting to Emmy Noether's scientific significance; he wanted to present these to the ministry in the hope of gaining permission for Noether to continue working in Göttingen. The names of mathematicians who responded to Hasse's request include H. Bohr, Ph. Furtwängler, O. Perron, T. Rella, J.A. Schouten, B. Segre, K. Shoda, C.L. Siegel, A. Speiser, T. Takagi, B.L. Van der Waerden, H. Weyl. See also [Roq10].

[441]Wolfgang Wichmann was a Noether doctoral student. He had initiated a petition to the ministry on Noether's behalf, which was signed by twelve students. They were members of what Noether in her previous letter had called "Noethergemeinschaft", who studied Hasse's Marburg lecture notes together in a seminar in Noether's apartment. The signatures include those of E. Bannow, E. Knauf, Ch. Tsen, W. Vorbeck, G. Dechamps, W. Wichmann, H. Davenport, H. Ulm, L. Schwarz, W. Brandt, D. Derry, Wei-Liang Chow. See also [Roq10]. – About Wichmann see also [Roq08]. The dissertation of Wichmann was inspired by Emmy Noether. Its main content was a simplified derivation of the functional equation of the zeta-function of a central simple algebra. This coincided with the functional equation found by Käte Hey [Hey29],[Zor33] up to a question of a sign that couldn't be determined this

way. (See [Lor08].) Wichmann's dissertation was completed in 1934 only, when Noether had already emigrated to the USA. Therefore, F.K. Schmidt formally served as referee for this dissertation.

[442]Section 3 was the Aryan paragraph of the new Nazi law; it excluded every "non-Aryan" (e.g., people with jewish background) from the Public Service in Germany.

[443]Noether refers to Hasse's letter to the curator of Göttingen University, where he announced sending him the recommendations on behalf of Noether by a number of internationally known mathematicians. It appears that Hasse had provided Noether with a copy of that letter.

[444]The name of the curator (manager) of Göttingen University was Valentiner. In his statement for the ministry he acknowledged Noether's scientific high qualification but he did not recommend her remaining at the university because of her Jewish background and her leftist political view.

[445]Hasse eventually was able to submit 14 reports from leading mathematicians about Emmy Noether, their names are cited above. (However there was one mathematician, Erhard Tornier, who vehemently was against Noether's remaining in Göttingen. In his letter to Hasse dated 23 September 1933 he threatened Hasse with *"very severe consequences"* should he submit the reports about Noether to the ministry. Nevertheless, Hasse did send the reports.) Hermann Weyl in his memorial speech for Emmy Noether at Bryn Mawr on April 26, 1935 reported: *"I suppose there could hardly have been in any other case such a pile of enthusiastic testimonials filed with the Ministerium as was sent in on her behalf. At that time we really fought; there was still hope left that the worst could be warded off. It was in vain."* [Wey35].

[446]Hasse had suggested in his letter to the curator of Göttingen University, that one might commission Noether with special lectures for a smaller group of gifted students – which, by the way, completely conformed to her actual activity up to then.

[447]The following sketch on class field theory seems to be a consequence of Noether's seminar about Hasse's Marburg lectures which she held in her apartment (see letter of May 5, 1933). She uses mainly the same notations as in Hasse's lecture notes, but her style is quite brief so that it is difficult to extract meaningful results from it. At the end of this paragraph she says that she wants to develop class field theory without using Furtwängler's Norm Theorem which was based on analytic considerations. But this had already been done in Hasse's Marburg lectures, following Chevalley and Herbrand. Obviously Noether was looking for another proof.

[448]This concerns an existence theorem for separating the rational numbers into congruence classes with given properties. Such an existence theorem was formulated and proved by Artin in 1927 as part of his proof of his Reciprocity

Law. Hasse formulated a generalization as Theorem (0.3) in [Has33b]. The proof was not elementary and depended on the Frobenius Density Theorem. An elementary proof for Artin's special case was given by Hasse in his Marburg Lectures [Has33d], which Noether refers to here; it is Hasse's Theorem 139. Chevalley in his thesis gave a proof that was elementary. Now Noether asks whether also the generalization as formulated in Hasse's Theorem (0.3) has an elementary proof. That is indeed the case. A proof was given by Van der Waerden, but *after* the date of this letter. There is a letter from Hasse to Davenport from October 15, 1933, in which Hasse recalls the problem of the Existence Theorem and writes: "*It may interest you that V. D. Waerden found a very simple elementary proof.*" It seems that Noether had posed the question not only to Hasse but also to Van Der Waerden, and the latter had found the solution. Van Der Waerden's proof was published in 1934 in Crelle's Journal [vdW34].

[449]Chevalley's thesis was published in Journ. Fac. Sci. Tokyo [Che33b]; perhaps the volume in question had not yet appeared at the time of this letter. However hectograph copies of this thesis circulated; we can assume that Hasse had a copy.

[450]Noether refers to Tsen's dissertation [Tse34]. This was never published in a mathematics journal but a preview version appeared in the Göttingen Nachrichten [Tse33]. His doctoral exam took place in 1934, when Noether was already in the USA. In Göttingen at that time, none of the fired math faculty could serve as an advisor. F.K. Schmidt, representing Hermann Weyl, was assigned to take over as formal referee for all current exams. – By the way, Tsen had mentioned Wichmann in his dissertation. Tsen reported that Wichmann had pointed out to him that, as a consequence of his theorem, over a real closed field every division algebra has index 2 and is therefore a (generalized) quaternion algebra. See the letter from July 21, 1933. §3.71 – Noether's suggestion that "perhaps" the Theorem of Tsen might hold for function fields in several variables is wrong: See the postcard from June 27, 1933. § 3.70 and the letter from July 21, 1933. §3.71. For more on Tsen see [Lor99].

[451]When Noether speaks of "our" transfer from cyclic to general, she is referring to the chain of reasoning developed by Hasse, Chevalley and herself. The earlier letter from June 3, 1932 discussed the transfer of the "Inverse Theorem" of local class field theory from cyclic fields to arbitrary abelian fields. Now the issue is to transfer Tsen's Theorem in a similar manner.

3.70 Postcard of June 27, 1933

Schwarz's Dissertation. Arithmetic Foundation of Class Field Theory. Tsen.

June 27, 1933

Lieber Herr Hasse!

I will work out the hypercomplex considerations a bit [more] although I don't yet know whether the method of crossing [with a cyclotomic field] works. The Artin lemma alone won't do it in any case.[452] That will take a little time because at the moment I must quickly read a dissertation (Schwarz). The official supervisor is Weyl[453]

Now I would like to know from you whether Chevalley really has *everything* arithmetic, or whether there is a basic misunderstanding on my part. According to his C.R.[454] (Feb. 32) he misses only the fact that for cyclotomic fields the class division A/H modulo the associated group is the same as the division into Artin classes (Théorème B for cyclotomic fields). But according to your manuscript, p. 167, top, line three,[455] elementary provable – at least as you make it – and is covered with your statement (0.4),[456] which you again list as an immediate consequence of the splitting laws in cyclotomic fields. Naturally I am also working with (0.4). Who is wrong?[457] I hope that my approach will serve to better clarify the entire arithmetic part of class field theory, not only the local theory.

The Theorem of Tsen is not valid for more variables, that is almost immediately clear. With one variable and real closed coefficients there are only "generalized quaternions".[458]

Herzl. Grüße, Ihre Emmy Noether

Notes

[452]Evidently Hasse had requested a more precise presentation of Noether's considerations in her previous letter of June 21, 1933. §3.69. Hasse had formulated "Artin's Lemma" in Theorem 139 of his Marburg Lectures [Has33d]. See the notes to Noether's letter of June 21, 1933. §3.69

[453]Emmy Noether had the authorization to serve on doctoral promotions taken away. The doctorand, Ludwig Schwarz, had finished a dissertation entitled "On the theory of non-commutative polynomial domains and quotient rings". In the promotion records of the faculty, Weyl was listed as first referee and Herglotz as second referee. Schwarz's dissertation was not published in a mathematics journal. We also couldn't find it in the Göttingen library.

[454]C.R. = Comptes Rendus Académie Paris. This concerns [Che32].

[455]The page number 167 refers to the original hectograph copy of Hasse's Marburg Lectures from the year 1932 [Has33d]. In the printed version [Has67] it is page 193, second paragraph, line 2.

[456]Theorem (0.4) is from Hasse's Annals paper dedicated to Emmy Noether on her 50-th birthday. It concerns one of the theorems listed in the introduction which were taken as known.

[457]We are presented with a misunderstanding, as Noether conjectured, namely, in the use of the term "elementary" for a proof. Noether and also Chevalley are searching for an "arithmetic" proof which would eliminate the need for any theorems about analytic functions. Hasse's "Theorem (0.4)" and also the corresponding statements on "page 167" refer to cyclotomic fields. The proof required, among other things, the Dirichlet theorem on the existence of prime numbers in an arithmetic progression, and at that time that could only be proved with the help of the analytic theory of L-series; the proofs are therefore not "arithmetic", as Noether actually wished, even if from certain points of view she could describe them as "elementary". – The fact that class field theory can be developed completely without analytic methods was announced later by Chevalley in a C.R. note [Che35]

[458]See the previous letter of June 21, 1933 § 3.69, and also the following letter of July 21, 1933. §3.71.

3.71   Letter of July 21, 1933

Hasse's Approach to Tsen's Theorem. Witt's Ideas on Class Field Theory in Function Fields. Witt's Dissertation. Arithmetic Proof of Bauer's Theorem via Deuring und Noether's Interpretation through Galois Modules. Relations Between Gauss Sums. DMV-Meeting in Würzburg. Document. Questionnaire.

Göttingen, July 21, 1933

Lieber Herr Hasse!

Last week I talked to Van der Waerden – who like Artin lectured about Lie groups – about your approach to Tsen's Theorem. He believes that it doesn't work that way, because you fix the auxiliary points, only their number is arbitrary, in order to come to *linear* equations despite the formation of integrals. He suspects that integrality of the solution requires certain constraints in the position of the points (there are continuously many divisors, but only countably many integral solutions), so that the points themselves still have to be taken as unknowns, which due to the integrals leads to transcendental equations. That in this case there are solutions follows from Tsen's result, but I admit that the counting with elimination theory is not nice.[459] Or are these thoughts lapsed and you or Chevalley have an arithmetic proof?[460] Davenport said something like that to Tsen.[461] His (Tsen's) note in the Göttinger Nachrichten will appear soon; he and others are very receptive for reprints or page proofs of your work, especially for Annals 104 and 107. (My copy, 107 with page proof, has already wandered.)[462]

Witt noted that the Tsen result can be understood as an analogue to your Cyclicity Theorem, for this allows the (weakened) formulation: "Over the field of *all* roots of unity every algebra splits".[463] And the constants in function fields are precisely the elements that everywhere have order zero, i.e. the analogue of the

roots of unity. From this remark it immediately follows that algebras have a cyclic representation when their center is an algebraic
function field in one variable with a *Galois field*[464] as coefficients.
For as splitting field one can take a *finite* extension of the Galois field, i.e., a "cyclic cyclotomic field". Witt hopes with this
approach that class field theory in the case of F.K. Schmidt can
be built up in a purely arithmetic-hypercomplex fashion.[465] He
can simply and explicitly write down algebras with prescribed invariants (whose invariant sum is zero); the Norm Theorem is still
missing.[466]

Witt will this week – he suddenly started to work and not
just to simplify[467] – complete his doctorate with a dissertation
"Riemann-Roch theorem in the hypercomplexes".[468]   I spoke
probably already in winter about the problem which I came up
with from the note by Zorn.[469] But I didn't think that F.K.
Schmidt[470] would simply transfer to "one-sided divisors" (the
composition of one-sided ideals from different places) which I
didn't have, which is actually the case, only that the genus G
of the division algebra differs from the g of the center by the
ramification orders, otherwise everything [holds] literally! And
for *any* coefficient domain. For a Galois field as coefficients the
functional equation of the hypercomplex ζ-function, theorem of
the split algebras etc. comes out. Here Witt has also included
the examples of the algebras.[471]

With the help of the constant splitting field, which exists according to Tsen, I hope that now one can also approach algebraic functions with rational coefficients – that particularly interests Artin.
As preparation I think first to discuss algebras where the center is
an algebraic function field with **p**-adic coefficients, like Tsen does
for real coefficients.[472] But with all that, I still haven't finished
writing up the recent hypercomplex class field theory considerations: I also believe that it is more about formal rather than material simplifications; replacing transcendental proofs with arithmetical, one must first of all wait for the Deuring-Bauer theorem
etc.,[473]; Chevalley surely has told you![474] By the way, couldn't
one arithmetically prove the existence of prime ideals of relative
degree *one* in advance? That follows at once from Bauer's the-

orem! That existence theorem says with respect to the above considerations: if you form the extension of the group J of the *ideals* (not ideal classes) with the Galois group G of K/k, then G becomes isomorphic and not just homomorphic to the generated automorphism group (since the prime ideals of relative degree one take n different values, [so] in some sense play the role of primitive elements).

I also see that I have not responded to your recent question – relations between Gauss sums. I only know that the specified multiplication can be interpreted as a composition of Galois modules in the p-th cyclotomic field; the "constants" $\psi^m(m)$ must therefore be related to factor systems; if you replace this with [the identity] one, it must have to do with compositions of representations of the cyclic group of degree $(p-1)$. Whether it's a well-known composition, I don't know. Can you do something with this? Or is that much known to you?[475]

Will you come to Würzburg by car, and lecture about these Riemann conjectures?[476] I'm not sure yet, should I go?[477] For now, I want to meet my siblings at the Baltic Sea: Address from July 31: Dierhagen near Ribnitz, Mecklenburg.

By the way, Weyl wanted to speak to the curator again because of the submission date of the affidavits: He began coincidentally with his affidavit, believed that the whole thing would soon be submitted. So you might get a message again. But then you would have to – that was also Weyl's opinion – in any case make copies which *you* keep. If that would be uncomfortable for you in Marburg it could of course also be done here.[478]

In a questionnaire that I have now received, I stated that Klein and Hilbert brought me to Göttingen in spring 1915 to stand in for a *Privatdozent* position. To conclude from that, that by August 1914 I had met the prerequisites, however, requires quite a bit of imaginary benevolence! Further, I have put on paper my socialist (before 1919–22, independent-soc.) party affiliation, until 1924; I never voted further to the left![479]

Nun herzliche Grüße! Ihre Emmy Noether.

Notes

[459]We do not know Hasse's approach to the proof of Tsen's theorem. But since Noether says that "in this case there are solutions", it seems likely that this refers to the second part of Tsen's Theorem, which concerns function fields over the *real* ground field $\mathbb{R}$ where division algebras do exist. But Tsen's result gives only the existence of non-trivial division algebras over such fields and, as Noether says, he used elimination theory in his proof which is "not nice". – It seems Hasse had tried an approach to determine the full Brauer group over a real function field, using abelian integrals. In this context we find it interesting that Witt also works with abelian integrals in his paper [Wit34c]. There Witt goes beyond Tsen and gives an explicit description of all division algebras over any function field in one variable with constant field $\mathbb{R}$.

[460]As in the previous postcard of June 27, 1933, Noether seems to use the modifier "arithmetic" as the opposite of "analytic". The present case is about getting to the result without the analytic theory of abelian integrals. An approach to an "arithmetic" proof in the sense of Emmy Noether was found only much later in the works of Geyer [Gey66], [Gey77]. The full "arithmetization", valid for any real closed constant field, makes use of the so-called semi-algebraic geometry, in which it is shown that the connected components of a real curve can be defined "arithmetically". See Claus Scheiderer's comments on [Wit34c], published in the Collected Works of Witt [Wit98], as well as the literature cited there.

[461]As already mentioned in the notes to the letter dated May 10, 1933, Davenport stayed in Göttingen in the summer semester of 1933 but often met up with Hasse in Marburg. Chevalley was in Marburg in the summer semester of 1933.

[462]Hasse's papers in volumes 104 and 107 were [Has31d] and [Has33b]. Noether had received a page proof copy of the latter paper, which she made available to her students, where it "wandered".

[463]That indeed follows from Hasse's results in his paper in Annalen 107 [Has33b]. However, many years later Hasse [Has49b] admitted a lapse in his proof and provided a correction. But in the meantime R. Brauer had already proved a sharper result [Bra47], so that Hasse's theorem had become outdated already.

[464]Here Noether means a field with finitely many elements, also in her later letters. This terminology had not yet been commonly used in the German mathematical literature; it seems that she had accepted it from the literature in English.

[465]When Noether writes "in the case of F.K. Schmidt" she means the algebraic function fields of one variable with finite field of constants. In this case, F.K. Schmidt had already started to develop class field theory, but not in an "arithmetic-hypercomplex" fashion, as Noether now wishes, since he used analytic L-functions. Moreover, F.K. Schmidt could only handle the case when the degrees of the abelian extensions are not divisible by the characteristic p of the fields. See [Sch31b]. (In a letter to Hasse dated December 6th, 1926, F. K. Schmidt points out the difficulties that arise when p divides the degree, and he writes: "*These exceptions do not interest me in my investigation, so I have excluded them from the start.*")

[466]Hasse used the same idea in his paper on cyclic algebraic function fields [Has34c], namely in the proof of the summation formula for algebras (and thus of Artin's law of reciprocity) with the help of the theorem of Tsen. Hasse thus followed the procedure in his paper [Has33b], dedicated to Emmy Noether on her 50th birthday; in the present case the situation is much easier because the constant extensions are unramified. From this Hasse derives the Norm Theorem. – For a *rational* function field as base field Hasse gives a second proof of the reciprocity law, which does not use the Theorem of Tsen. Later this second proof idea was generalized to function fields of arbitrary genus, by H.L. Schmid and Witt among others. See also [Roq01].

[467]Already as a young student, Witt was considered to be a master at finding particularly simple proofs. His first achievement was a proof of commutativity of finite division algebras, which took up less than one side of a single sheet of paper [Wit31].

[468]Witt himself says in his inaugural address in front of the Göttinger Academy (1983): "*I was deeply impressed in 1932 by the three famous lectures by Artin about Class Field Theory. I then spent the next holidays in Hamburg intensively studying Class Field Theory for number fields. In the years that followed, it was my goal to transfer Class Field Theory to function fields. The first step led to my doctorate in 1933...*" See [Wit98]. – Ina Kersten reports in her Witt biography [Ker00] that Witt chose his dissertation topic himself, namely the working out of a general problem area posed by Emmy Noether. He started work on July 1, 1933 and delivered the finished dissertation a week later, on July 7, 1933. Noether's task required the development of the methods of Hey's dissertation (with Zorn's supplement) for algebraic function fields with finite constant fields. This was in line with his declared goal of transferring the theorems of class field theory to the case of function fields.

[469]Zorn's note in the Hamburger Abhandlungen [Zor33] showed that after certain corrections and additions, the Hamburg dissertation of Käte Hey [Hey29] could be used for a proof of the Local-Global Principle for algebras over number fields. It is an analytic proof, using the zeta function of

an algebra. Zorn presented his paper to the editors of the Hamburger Abhandlungen in January 1933. Apparently Noether was aware of Zorn's result
before it went to press because she says that she had already spoken about
this problem in her lecture in the winter semester.

[470]This refers to F.K. Schmidt's proof of the Riemann-Roch theorem for
function fields in [Sch31a].

[471]Witt's dissertation was published in the *Mathematische Annalen*
[Wit34a].

[472]Algebras over function fields of one variable with real constant field $\mathbb{R}$
were systematically studied in [Wit34c]. The case of a $\mathfrak{p}$-adic ground field is
much more difficult.

[473]Deuring's *arithmetic* proof of Bauer's theorem is based on partial statements of class field theory and was published in Crelle's Journal[Deu35b].
Bauer's theorem says, among other things, that a Galois extension field of
an algebraic number field is uniquely determined by the set of prime divisors
that split in it. Bauer's original proof dates back to 1916 and uses tools from
analytic number theory. See also the postcard from September 13, 1933.3.73

[474]Chevalley stayed in Marburg with Hasse in the summer semester 1933.
From there he visited Emmy Noether in Göttingen and heard the latest
mathematical news.

[475]At this time Hasse was working on the analogue of the Riemann conjecture in function fields with finite constant field. Hasse succeeded in proving
this for elliptic function fields in February 1933. Now together with Davenport, he studied certain function fields of higher genus; today they are called
"function fields of Davenport-Hasse type". Cf. [DH34]. For these function
fields it is possible to represent the zeros of the associated zeta functions by
Gauss sums. For this purpose it was necessary to generalize the theory of
Gauss sums in a suitable way. Hasse interpreted the so-called Jacobi sums as
factor systems defined by Gauss sums, and that seems to interest Noether.
See also the relevant note on the following letter dated September 7, 1933. 3.72

[476]In September 1933, the annual meeting of the DMV was held in
Würzburg. Hasse spoke there about his proof of the Riemann conjecture for
elliptic function fields.

[477]There was a discussion among the members of the DMV whether one
should participate there as usual, including those members who were fired
by the Nazis from their positions, in order to demonstrate that the DMV is
a society of scientists, not politically orientated. Or should one abstain from
participation as a protest against those Nazi laws? Finally, Noether decided
not to go to Würzburg.

[478]Re. affidavits, see the letter dated June 21, 1933.3.69

[479]To eliminate faculty with Jewish backgrounds from German universities, the Nazi government required every professor to complete a questionnaire asking for race, religion, political affiliations and similar questions about parents and grandparents. Apparently the Nazi law contained a clause excepting those people who had fought in the army or held a position in public service during world war I. Noether seems skeptical that being called to Göttingen by Klein and Hilbert during the war would meet this prerequisite. And indeed she was fired from Göttingen University on September 2, 1933. Copies of the documents about Emmy Noether during this time can be found at my homepage:

https:/www.mathi.uni-heidelberg.de/roquette/gutachten/noether-gutachten.htm

3.72   Letter of September 07, 1933

Again: Recommendations. Noether's Plans for Oxford, Bryn Mawr and Princeton. Noether does not go to Würzburg. Again Gauss Sums.
F. K. Schmidt. Schilling. Deuring Returns from the USA.

Göttingen, September 6 and 7, 1933

Lieber Herr Hasse!

Only after my return last week did I – surely your stay in Oberstdorf is over? – hear that when turning in the recommendations you also had to submit a letter to help. So really thank you very much for all your trouble! If not for now, maybe the reports will help for later! And it seems right to me that they are now available! [480]

You have heard from Davenport that I want to go to Oxford for a term after Christmas. Meanwhile, I received an invitation for a research professorship at Bryn Mawr for one year (1933/34), which I have accepted for the following 34/35. I don't have an answer yet, but I think the postponement – I can't be in England and America at the same time – will cause no difficulties. The scholarship comes jointly from Rockefeller and the Committee "In Aid of Displaced German Scholars". Bryn Mawr is a women's college, by the way, but as Veblen wrote later, it is the best of these; and is also so close to Princeton that I can visit often. In addition, we have to first wait for the decisions here.[481] (Courant has an invitation for the winter in Cambridge, Weyl gives lectures in America until early December; it is not yet known whether Landau will read his course.)

I hope that the representation of F.K. Schmidt can be realized.[482], and that you too will once again come to lecture in the Math. Society – but then when I'm there! After a suggestion that seemed reasonable to me, from Blaschke whom I met with Rademacher on the sea, I likely will not attend Würzburg.[483] Blaschke said it was primarily important that the Mathematicical Society preserves its purely scientific, neutral character, and that other questions not come up at all. This could possibly be made more difficult by my attendence this year.[484] This seems to me more feasible than Rademacher's suggestion that all those put on leave should definitely be given invitations [to Würzburg] since this is a meeting of mathematicians and not professors; in particular since probably most of them wouldn't come and an unnecessary question of principle would not arise. Blaschke thought about corresponding with you about all these questions, by the way; I don't know if he did it! I think he is now in Geneva.

I've looked again at your Gauss sums. What I wrote the other day gives just the "algebraic" [aspect] of [the] relations; an arithmetic consideration must then still determine the factor system $\psi^m(m)$.

I use from the theory of Galois modules: If K/k is abelian, taken as hypercomplex over k, I form the extension from k to $\overline{k}$, where $\overline{k}$ arises from k by adjunction of all characters of the group.[485] Then $K_{\overline{k}}$ becomes the sum of n Galois modules of rank one, which correspond one-to-one to the n representations.[486] Each representation arises from *one* basis element, image of the idempotent of the group ring; the basis elements are determined up to factors in $\overline{k}$. Let $\tau(\chi_1)$, ..., $\tau(\chi_n)$ be fixed basis elements of the n Galois modules, i.e. $\tau^S(\chi_i) = \chi_i(S) \cdot \tau(\chi_i)$. It follows that the product $\tau(\chi_i)\tau(\chi_k)$ corresponds to the character $\chi_i\chi_k$. More precisely: $[\tau(\chi_i)\tau(\chi_k)]^S = \tau^S(\chi_i)\tau^S(\chi_k) = \chi_i(S)\chi_k(S)\tau(\chi_i)\tau(\chi_k)$. The one-one correspondence of Galois modules and representations gives: $\tau(\chi_i)\tau(\chi_k) = c_{ik}\tau(\chi_i\chi_k)$, where c_{ik} is an element from $\overline{k}$. With the *chosen* basis the c_{ik} are *uniquely* determined; on the other hand the class of c_{ik} with the general element $\overline{c}_{ik} = \frac{r_i r_k}{r_{ik}} c_{ik}$, where the r are again in $\overline{k}$, is determined by the field *invariant*.

(Special case: determination of a pure equation modulo nth powers in the cyclic [case]; these are the invariants that Deuring also wanted to use in the non-abelian case of field-normal form.[487]) The repetition of this type of reasoning results in: " If a character χ has two decompositions: $\chi = \psi_1 \cdots \psi_s = \varphi_1 \cdots \varphi_t$, the corresponding basis elements satisfy: $\tau(\psi_1) \cdots \tau(\psi_s) = c \cdot \tau(\varphi_1) \cdots \tau(\varphi_t)$ with c in $\overline{k}$, then with τ fixed c is uniquely determined, and is determined by the field up to associated invariants."

Such a decomposition is now present in your relations:

$$\psi \cdot (\chi\psi) \cdots (\chi^{m-1}\psi) = (\psi^m) \cdot \chi \cdots \chi^{m-1}.$$

The above consideration results in:[488] The formula

$$\prod_{\mu \bmod m} G(\chi^\mu\psi) = c \cdot G(\psi^m) \prod_{\mu \not\equiv 0 \bmod m} G(\chi^\mu),$$

holds in general; so also for K the field of the p-th roots of unity, where now $G(\chi)$ is the chosen fixed, natural basis element of the Gauss sum ($G(\chi)$ is the basis element for the representation χ^{-1}; but the character decomposition also applies in the form $\psi^{-1}(\chi^{-1}\psi^{-1}) \cdots = (\psi^{-m}) \cdot \chi^{-1} \cdots$). Since the *basis is declared to be arithmetic* – the integer group ring is mapped operator-isomorphically precisely on the maximal order of K – naturally the *factor system declared to be arithmetic*, which you already wrote the last postcard. Because of this arithmetic distinction the determination of c must use arithmetic properties of K. The decomposition of the root numbers, from which you took the relations, as Davenport told me, appear naturally, at least for $c = \psi^m(m)$.

Your considerations with the crossed product is only another interpretation of forming the product of the Galois modules or of its basis elements, can only yield the "algebraic" part; that's also in your postcard. It seems to me that the different interpretations of the Galois modules is of independent interest.

Schilling[489] reported to me at the end of July about Marburg. I will now write up the whole hypercomplex interpretation of class field theory – for the time being there is nothing left.[490] Deuring arrives in Bremen today – he still had a look at America – I will tell him to write down his arithmetic proof of Bauer's theorem (if it is correct) soon. Those were Schilling's orders.[491]

Herzliche Grüße, Ihre Emmy Noether.

How is the Riemann Conjecture for function fields going?

Notes

[480]See the previous letter from July 21, 1933 § 3.71, as well as the earlier one dated June 21, 1933. § 3.70.

[481]Emmy Noether explains her plans for the next two years because she was dismissed from Göttingen. What strikes us in her letters from this time is the noticeable lack of any grievance or complaint related to her dismissal, although this will change her life in a drastic way, as she was well aware. Her optimism shows through again and again, that everything is just temporary and soon "normal" conditions would occur again. And above all: her letters are still full of mathematical ideas.

[482]F.K. Schmidt, at that time *Privatdozent* in Erlangen, was called to Göttingen at the proposal of Courant in order to represent the professorship of Weyl. He worked in this capacity in Göttingen in the years 1933/34 and was then appointed full professor in Jena.

[483]The place of the annual conference of the German Mathematicians Society. See also the previous letter dated July 21,1933. § 3.71.

[484]We note that during the last 25 years Emmy Noether had attended almost every year the meeting of the German Mathematical Society!

[485]Meaning the adjunction of the values of all characters of the Galois group of $K|k$.

[486]Here n means the field degree: $n = [K : k]$.

[487]In several publications in the years 1948–1950 Hasse examined these normal forms more closely in both the abelian case as well as for any Galois extension.

[488]The following formula was discovered by Hasse in June 1933 after Davenport used it in the special case $m = 2$. That comes from Hasse's correspondence with Davenport. The latter spent the summer semester of 1933 in Göttingen and visited Hasse in Marburg often on the weekends; but there are also letters from this period between Marburg and Göttingen, which show that Hasse and Davenport intensively investigated Gauss sums, since these lead to a representation of the zeros of the zeta function of a function field with the equation $x^m + y^n = 1$. The formula deals with the following situation: χ and ψ are characters of the Galois group of the field of p-th roots of unity (p an odd prime number); m and n are the orders of these characters. Noether writes $G(\chi)$ for the Gauss sum of χ (Hasse calls it $\tau(\chi)$ – however with a different normalization, so that $\tau(\chi)$ becomes a basis element for the representation χ). As Noether explains, the given formula holds for a "certain" constant c in the field of $(p-1)$th roots of unity. This follows solely from the "algebraic" considerations that Noether discusses. By determining the prime decomposition of the Gauss sums, Hasse first proved that c is an m-th root of unity. He conjectured (and later proved) that $c = \psi^m(m)$. Apparently he had communicated this conjecture to Emmy Noether, since this was addressed in her letter.

The proof can finally be found in the joint work of Hasse and Davenport in Crelle's Journal [DH34]; see the formulas (0.9_1) and (0.9_2) found there. There the situation is even more general, where χ, ψ are multiplicative characters of any finite field $\mathbb{F}_{p^n}$ (and (not only of $\mathbb{F}_p$).

[489]See the letter dated May 10, 1933. § 3.68.

[490]Noether is referring to the statements in her letter dated June 21, 1933 3.69. Hasse had asked her to put her ideas on paper.

[491]Apparently Schilling had inquired about the status of Deuring's manuscript, which Noether had mentioned in her letter dated July 21, 1933. Deuring's paper [Deu35b] was accepted by Hasse for publication in Crelle's Journal.

3.73   Postcard of September 13, 1933

Withdrawal of venia legendi for Noether.   Deuring's Proof of Bauer's Theorem.

September 13, 1933

Lieber Herr Hasse!

Today came the venia withdrawal after § 3[492]; nevertheless, the reports can be used later and still have value! Thanks again! Deuring will come to Würzburg and can tell you himself about his new considerations. Bauer's theorem is correct; the proof is surprisingly simple. He transfers Theorem 89 of Hilbert's Zahlbericht to ray classes; from this it follows that for all relatively cyclic fields K/k with $h \geq n$, there are prime ideals in k of absolute first degree which do not remain of first degree in K.[493] Then transition to galois and arbitrary fields by decomposition groups and usual conclusions.

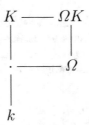

(If K is not a subfield of Ω, where K/k is galois, then $(\Omega K : \Omega) > 1$, and $\Omega K/\Omega$ galois: so in Ω there is a prime ideal of absolute degree 1 that does not split in ΩK, therefore such a $\mathfrak{p}$ in k that in Ω, not in K, splits off a first degree factor). The arithmetic proof that the relative cyclotomic fields [are] class fields doesn't work at the moment; but it doesn't look so improbable anymore.[494]

Best regards, your Emmy Noether.

Notes

[492]This was the final withdrawal of the *venia legendi*, that Emmy Noether had received through her habilitation in 1919 (for the latter see [Tol], vol. 2). Now Noether wasn't allowed to hold lectures any more in Göttingen.

[493]Theorem 89 states that the ideal class group (in the absolute sense) is generated by the prime ideals of the first degree. – The so-called second inequality $h \geq n$ of class field theory could already be proven at that time without analytic methods; hence Deuring's proof is to be seen as "arithmetic".

[494]Here Noether is again concerned with finding an arithmetic proof of the reciprocity law. See the notes to the letter of June 27, 1933. § 3.70.

3.74 Letter of March 6, 1934

From Bryn Mawr. Thanks for Separata. Publication plans. Visit in Göttingen. Teaching contract for Deuring? Report on Bryn Mawr and Princeton. Olga Taussky. Albert. Vandiver. Brinkmann on the integrality of Artin's L-Functions.

Bryn Mawr, Pa. U.S.A., March 6, 1934 [495]

Lieber Herr Hasse!

Finally I am getting around to thanking you for sending the reprints, the Higher Algebra [volume]I and the card from Kassel.[496] Namely, I wanted also to send the enclosed sketch of class field theory in the galois case and that took time.[497] Whether one really can regard things as a decomposition law depends on whether the given characterization is sufficient for an existence theorem; and I know nothing at all about that. I want to first understand your small note on explicit construction in the cyclic case; perhaps they, together with Deuring's reflections on galois modules, give the key – Scholz also exists![498] But that is so uncertain that I probably first of all will work out and publish what is already existing – but I hope that the sketch makes it roughly understandable.

The considerations are heuristic [and] all hyper-complex; it is a matter of a division into classes according to the invariants with respect to the individual decomposition fields, to which the center relations come (i.e. in the abelian case only direct products of ideal algebras will be considered, as I told you in Marburg last year). But it was more convenient and above all more precise to give independent definitions: the value range corresponds to the group of invariants. It is essential that the unit class of the Frobenius classes in the modified sense becomes the assigned group; instead of the full ray the partial ray appears (p. 3). The arrangement is exactly as in Chevalley. (Carbon copies have been sent to Chevalley and Deuring, [could you] possibly forward your copy to Artin? The machine couldn't handle more. I have retained a very bad copy for myself.) [499]

From the Göttingen students – Witt, Bannow, Tsen – I have heard that you will probably be in Göttingen in the next semester? Is that correct? [500] I would like it very much, since in the meantime I intend to come to Göttingen for a few weeks toward the beginning of June; but I don't quite think that it's happening so quickly for you. In any case, I wanted to write to you about my teaching assignment in the fall, which I have already discussed with F.K. Schmidt. Don't you think that it is possible that Deuring gets it? A teaching assignment is basically what is most important to him, and I would wish very much for it to go to him, also the collaboration with you in which both parties would gain a lot! [501] He just wrote to me that he has thought about the algebraization of "complex multiplication" – without having quite penetrated it – but which is related to the question, given algebraic function fields with coefficients in an algebraic number field: which divisor classes have finite order? [502] Aren't there also relationships to your Riemann conjecture, which, as you write, you now have purely algebraically in the elliptic case.[503] And did you get any further?

The people here are all very accommodating and have a natural cordiality that allows you to be familiar with them even if it doesn't go very deep. One is invited frequently; I have also met all sorts of interesting people not college related. In the meantime I'm doing a seminar here with three "girls" – they are rarely called students – and a lecturer, and right now they are reading Van der Waerden vol. I with enthusiasm, an enthusiasm that includes working through all the exercises – certainly not required by me. On occasion I slip in some Hecke, beginning chapter.[504] For next year there will be, typically American, an "Emmy Noether Fellowship", that probably will be shared by a student of MacDuffee and one of Manning-Blichfeldt-Dickson; the former seems to have some level. Aside from that Miss Taussky will probably come with a Bryn Mawr Scholarship here;[505] she had already applied for it last year, but it was not awarded due to lack of funds – previously five such foreign scholarships were awarded, the depression shows itself everywhere! Finally, a student of Ore has applied for a National Research Fellowship to come here.

On the other hand, since February I have begun a weekly lecture at Princeton – at the Institute and not at the "Men's-University" which does not allow anything feminine, while Bryn Mawr has more male faculty than female, and only in the students is it exclusive. I have started with representation modules, groups with operators; in this winter for the first time, Princeton will be treated algebraically but at once thoroughly. Weyl also lectures on representation theory, however he wants to go to continuous groups. Before Christmas, Albert, on a "leave of absence" there, has spoken about something hypercomplex according to Dickson, along with his "Riemann matrices".[506] Vandiver, also on "leave of absence", lectures on number theory, for the first time in human memory at Princeton.[507] And von Neumann[508] – after a survey to the mathematics club by me about class field theory – ordered twelve copies of Chevalley[509] (Bryn Mawr should also get something out of it!) That's how I found out also that your manuscript will be translated into English, now hopefully in enough copies – in the Fall I had already pressured people for that.[510] I have hard-working research fellows as listeners, alongside Albert and Vandiver, but I notice that I have to be careful; they are used to doing explicit arithmetic, and I've already driven some away!

Taken together, University and Flexner Institute[511] have more than sixty professors and such who want to become one, although Princeton tries to pull everything in, the many research fellows are a sign of academic unemployment.

Brinkmann, at Swarthmore College, a half hour drive from here, has proved the Artin conjecture for a number of groups – I think congruence groups and those that can be represented fully symmetrically – that the L-functions are entire, through explicit calculation of the characters. Surely Artin didn't calculate this far? Anyway, I told him to write things down and publish; think of Math Annals if it's new and not too complicated.[512]

Nun herzliche Grüße Ihre Emmy Noether

Notes

[495]Since the last letter from Göttingen six months have passed. In the meantime Emmy Noether had to leave Göttingen and she had accepted an invitation as a visiting professor at Bryn Mawr, a girl's college near Philadelphia in the USA. The main impetus for this assignment came from Solomon Lefschetz who had met Noether in Göttingen and in Zürich. See the archive material of C. Kimberling on Emmy Noether, which is now contained in the University Library in Göttingen.

[496]Kassel was the home town of Hasse where he grew up and where his parents lived.

[497]We have not found this sketch in the Hasse estate. It was probably the elaboration which Noether had announced in the postcard of June 27, 1933, aiming at a generalization of class field theory for non-abelian field extensions. The main idea was to characterize galois extensions $K|k$ by the subgroup of the Brauer group of k which contains those algebras which split in K, respectively the invariants of those algebras.

[498]Hasse had published a paper in the *Mathematische Annalen* on the explicit construction of cyclic class fields [Has33c]. This paper was mentioned by Noether in her letters dated March 3 and March 23, 1933. – Arnold Scholz, who was in correspondence with Hasse, had published a series of papers on class field theory in the 1930s, mostly with explicit constructions and algorithmic foundation. See [LR16].

[499]Apparently these things did not lead to anything precise and were not published.

[500]Finally, Hasse started his professorship in Göttingen in July 1933, after heavy politically motivated problems. See [Sch87].

[501]Deuring did *not* get the teaching position that became free through Emmy Noether's dismissal. At Hasse's instigation, Deuring later earned his habilitation in Göttingen but, for political reasons, he did not receive the *venia legendi* there (i.e., the official permission to give lecture courses at the university of Göttingen), although Hasse had applied for it for him. See [Roq18], Chap. 9.

[502]Deuring's paper "Algebraic foundation of complex multiplication" appeared in the *Hamburger Abhandlungen* in 1949: [Deu49]. It can be gathered from the correspondence between Hasse and Deuring that the basic idea was finished already in 1937. Here we learn that Deuring had already started dealing with this question in 1933.

[503]In the course of the year 1933 Hasse was able to replace his first proof of the Riemann conjecture for function fields in the elliptic case (which referred to complex multiplication and class field theory, with a new proof that worked purely algebraically in characteristic $p > 0$. (See [Roq18].) Apparently Noether is alluding to this new proof. In fact, in Hasse's theory the structure of the divisor class group of finite order plays an important role.

[504]Hecke's "Lectures on the theory of algebraic numbers" [Hec23] was not translated into English until 1981; so Noether must have used the German edition as a basis, i.e. she had to require that her students had a knowledge of German. And the same with the algebra book by Van der Waerden [vdW30].

[505]Olga Taussky (the later Mrs. Taussky-Todd) did indeed come with a scholarship from Bryn Mawr. In 1981 she reported on her personal memories of this time with Emmy Noether in a commemorative volume [TT81]

[506]Some years earlier, in 1931, Albert made a significant contribution to the proof of the Brauer-Hasse-Noether theorem about the cyclicity of algebras. See the last paragraph in Noether's letter dated November 22, 1931 § 3.39. See also [Roq05b]. – In 1937 Albert published a textbook "Modern higher algebra", which, at least in the title, claims to be an English equivalent of Van der Waerden's book "Modern Algebra" [vdW30]. It is to be assumed that Noether's lectures at Princeton, which Albert attended, had a significant influence on the design of this book – next to the correspondence between Albert and Hasse, which lasted until 1935.

[507]For a number of years Hasse exchanged reprints with Vandiver (who worked in Austin, Texas) and there was some mathematical correspondence between them. Vandiver visited Hasse at least twice in Germany, once in Halle and once in Marburg.

[508]John von Neumann was one of the first six professors at the Institute for Advanced Study in Princeton.

[509]This is the thesis of Chevalley [Che33b], in which the class field theory is developed *ab ovo* and could be used well as a "textbook". This paper had appeared in Journal Fac. science Tokyo and had 111 pages.

[510]This refers to the elaboration of Hasse's Marburg lecture 1932 on class field theory [Has33d]. This had been distributed in the spring of 1933, but only to such interested parties who had previously ordered a copy. When Noether writes that she had "already pressured people in the Fall" for a desired translation, then that was in Oxford, England when Noether was there. From the correspondence Hasse-Mordell we know that actually a translation plan existed. However, this was later abandoned. See also the letter of April 5, 1932, as well as the postcard dated November 25, 1932.

[511]Flexner Institute = Institute for Advanced Study.

[512]This is about Heinrich W. Brinkmann, who from 1933–1969 taught at Swarthmore College, but apparently during this time published nothing. In any case, we haven't found any paper by Brinkmann on this topic. (Noether mistakenly writes "Brinckmann" instead of "Brinkmann".)

3.75   Letter of April 26, 1934

Congratulations to H. on the call to Göttingen. Class field construction and Riemann conjecture in function fields. Abelian function fields. Riemann matrices (Albert). Lefschetz. Irreduclble galois representations and class fields. Deuring's habilitation. Finland?

Bryn Mawr College (until May 18), April 26, 1934

Lieber Herr Hasse,

These days I heard from F.K. Schmidt that you have now really received the call to Göttingen, and I would like to tell you how happy I am about it! Now Göttingen remains the center![513] Congratulations!

But actually I wish you even better luck with your latest mathematics: Hilbert's problem of class field construction and the Riemann conjecture in function fields at the same time, that's a lot![514]

As far as I can see, your meromorphism ring[515] is closely related to the algebras belonging to Riemann matrices that Albert has now characterized (most readable in the latest issue of the Annals of Mathematics, Jan[uary] 1934[516]). That would settle the question whether the ring can be non-commutative and whether transcendental meromorphisms exist; the latter of course, only if Albert really has the full ring and not just the subring of all algebraic meromorphisms.[517] I will ask Albert about that tomorrow. I have also seen in this paper that the geometers have long been using the p-dimensional variety corresponding to your field of abelian functions;[518] of course all of this without any idea from number theory.[519] Lefschetz and Albert were very surprised when I said in a lecture (at the meeting in New York), that it seemed to me that the material for the solution of Hilbert's problem would lie in extending complex multiplication through abelian functions. Days later your letter arrived [stating that] you really did it![520] To be sure, with a lot of deep methods!

When are you going to Helsingfors – and to the Finnish lakes?[521] And what about Göttingen in the summer? I'm counting on you being there some of the time, even if not all the time! I want to come in the first days of June, via Hamburg, where I will probably stay two days. There are hardly any Easter holidays here; therefore the Institute in Princeton closes on May 1st, here the lectures [end] in mid-May. Then there are exams here, where I am not involved, and a ceremonial procession at the end, with three different graduate gowns, B.A., M.A. and Ph.D.

By the way, concerning the class division I recently sent,[522] in the meantime it has become more and more clear to me that it is just as much a transmission of everything "abelian" as when Artin dealt with the transmission of everything "cyclic". So one cannot approach the existence theorem, if only because the local classification – i.e. for solvable fields at the ramification points – completely fails. It seems to me that one must first have clarity here, i.e. not just an overview of the relations in successively cyclic [extensions] via composition series, but in relation to the irreducible representations, i.e. to the galois modules,[523] and thus also to the Artin conductor. But at the moment I haven't made any progress.

On behalf of Deuring, I am particularly pleased that you are coming [to Göttingen]. I hope that the [Deuring's] habilitation will go quickly; unfortunately, he missed [the chance] to do that before [going to] America, and it seems to drag out in Leipzig with the new regulations.[524] For his hab[ilitation] manuscript he has worked out his results from last summer on the Riemann hypothesis for systems of quadratic imaginary fields.

Nun herzliche Grüße Ihre Emmy Noether.

Notes

[513]In April 1934 Hasse had received a call to Göttingen, which he was inclined to accept, since he felt obligated to restore the importance of Göttingen as an international center of mathematics as far as possible after the painful losses following the seizure of power by the Nazis. However, as a result of the turbulent events in Göttingen it took until the beginning of July 1934 for Hasse to be able to accept and actually take up his appointment. And despite Hasse's efforts, Noether's desire that "Göttingen remain the center" could not be realized in light of the political development in Germany in the following years. Already a year later, on April 18, 1935, Hasse wrote to Toeplitz [translated]: *"What rather depresses me is the fact that, on the one hand, I carry the responsibility toward the mathematical world for the reconstruction of Göttingen to a mathematical place of importance, but on the other hand by the existing university policy regulations, I have no influence on almost every crucial decision on the staffing. This applies not only to the filling of professorships, but applies in the same way to teaching assignments, assistants and student worker [Hilfsassistent] positions."*

[514]We do not know what F.K. Schmidt (or Hasse himself) had written to Emmy Noether. According to her comments in this letter it was probably a

program for algebraizing the theory of abelian function fields (characteristic p included), aiming at a proof of the Riemann conjecture for function fields of any genus. Already in his first publication on the Riemann hypothesis in elliptic function fields [Has33a] Hasse had suggested such a program. However, this program was never carried out. Note that Noether does not congratulate Hasse on his results, rather she "wishes him luck for his mathematics", i.e. successful further work on his project. (Later it turned out that significant difficulties arose in the detailed implementation. As we now know, Hasse did not reach his goal of finding a proof for the Riemann conjecture in function fields of arbitrary genus. André Weil anticipated him in this. See [Roq18].)

As far as we know Hasse never worked seriously on the "Hilbert Problem" of class field construction mentioned by Emmy Noether (the 12th Hilbert problem). However, at that time there was generally the expectation that in the course of the algebraization of the theory of abelian function fields (which Hasse aimed for) light would also be shed on Hilbert's 12th problem. Maybe Hasse informed her of this connection. (Today we know that this applies not for every number field, but only for the so-called CM-fields, i.e. those that occur in in connection with complex multiplication for abelian varieties.)

[515]In today's terminology this is the endomorphism ring of an abelian variety.

[516][Alb34].

[517]At this point in time Hasse had already discovered that even for elliptic curves in characteristic $p > 0$ non-commutative meromorphism rings could occur. Noether probably means the question of whether non-commutativity can occur in characteristic 0 and in particular over number fields, in the case of abelian varieties of higher dimension.

[518]Here, Emmy Noether inserts a footnote that reads as follows: "*These written explanations helped a lot when reading the corrections!*" This also indicates that Hasse had written to her a sketch of the algebraic theory of abelian function fields, in the way which he imagined. – But it is not clear what Noether means when she speaks of "corrections". Perhaps she means galley proofs of the paper of Albert [Alb34] mentioned above. Or is it the galley proof of Hasse's paper in the Hamburger Abhandlungen on the Riemann conjecture in the elliptic case? [Has34a].

[519]Here p does not mean the characteristic, but the genus of the respective function field (or curve).

[520]For this, see also the preceding footnotes. Here, probably "you did it" should be understood in the sense of "saw a possible way to do it".

[521]This playful allusion to Finland seems to be in response to learning that Hasse had invited the Finnish mathematician Nevanlinna and his pupil

Ahlfors to lecture in Marburg. In fact, these lectures took place in May 1934. (In a letter to Davenport on May 27, 1934 Hasse said he was very impressed by the lectures.) We do not know when Hasse contacted the Finnish function-theoretical school. Nevanlinna had visited Göttingen several times in the 1920s and met Emmy Noether there too; maybe Hasse got to know Nevanlinna on one of those occasions. Or maybe the contact was established through Egon Ullrich, who habilitated with Hasse in Marburg. Ullrich had studied in Helsinki in 1927 and there met Nevanlinna, with whom he formed a lifelong friendship. Hasse's correspondence with Nevanlinna began in 1932 and continued unabated.

After moving to Göttingen in July 1934, Hasse saw to it that the theory of complex functions was again represented, and so he took Ullrich with him as senior assistant. A trip for Hasse to visit Finland (" to the Finnish lakes") was planned for September 1934, but Hasse couldn't go because of the unstable situation at Göttingen. He did not make this trip until October, 1938. Hasse succeeded in getting Nevanlinna to come to Göttingen for a year as a visiting professor in 1936. See the letter from April 7, 1935.

[522]see the previous letter dated March 6, 1934.

[523]Noether's comment here that the irreducible representations (meaning representations of the galois group) are indispensable for non-abelian class field theory, calls to mind the Langlands program, which was conceived much later.

[524]On Deuring's habilitation see the comments on the previous letter dated March 6, 1934.

3.76   Letter of June 21, 1934

Letter from Göttingen.   Report on N.'s lecture in Hamburg on galois class field theory. Witt. Tornier.

Göttingen, June 21, 1934

Lieber Herr Hasse,

I've been back here for 14 days, of course I was very sad not to find you; after the latest news from Hamburg I had half expected that![525] But at the moment everything seems to look favorable again!

I'm sorry that I probably can't invite you to my place now;[526] and in the Fall I'll give up the apartment and take books and the good stuff from my things with me.[527] I was assured over there that I can expect [an] extension; whether in Bryn Mawr or elsewhere, probably they do not know yet. The "appointments" in America are only ever given for two years, even in the regular case where there are of course customary rights.

Aren't you coming some time to visit your parents in Bad Soden? Because it's better if I don't visit you in Marburg either?

Many greetings from the Artins, where I have stayed for a few days; in the new apartment in Langenhorn (Hainfelderstr. 9), which is really like a summer resort. There a second child is expected one of these days, as you probably know. Artin did not write to you on the occasion of your apointment in Göttingen; he apologizes for this by his general non-writing.

In Hamburg I have lectured as "Noether–Amerika" in the Artin seminar, extensively about the division into classes in the case of galois fields. Artin, who once thought about something similar, without getting through it all, is as skeptical as I am in regards to the existence theorem. Chevalley who was there for a week had one hypercomplex suggestion, but for prime powers it does not agree with the Artin L-series. One maps the ideals in K ring-homomorphically to the integers of the center of the group ring, by first extending the Frobenius substitution $\mathfrak{p} \to \langle \mathcal{S} \rangle$ [528] to a multiplicative homomorphism $\mathfrak{p}_1^{\rho_1} \cdots \mathfrak{p}_r^{\rho_r} \longrightarrow \langle \mathcal{S}_1 \rangle^{\rho_1} \cdots \langle \mathcal{S}_r \rangle^{\rho_r}$, with multiplication understood in the group ring, and then [uses] the linear expressibility of the $\langle \mathcal{S} \rangle$ powers by the classes $\langle \mathcal{S} \rangle$ themselves "transferred" to the ideals, i.e. by this addition of the ideals. The unit class of the ideals would then be all those that correspond to $\langle \mathcal{E} \rangle$ in the group ring.

In fact, however, Artin does not assign the $\mathfrak{p}$-powers from the L-series to the powers $\langle \mathcal{S} \rangle^\rho$, but rather to the class of $\mathcal{S}^\rho$: $\langle \mathcal{S}^\rho \rangle$

– which agrees with my class division – so that [if] I split off
the decomposition group Chevalley comes out, which I also don't
rightly believe.[529]

Witt now believes, with your Riemann conjecture for any function field, that he is able to prove the functional equation of the
L-series and also some more; he already had such hypothetical
approaches.[530]   You will [already] know that Witt has found a
counterexample to the theorem about split algebras – ground field
$\mathsf{P}(x)$ with P rational numbers.[531]

Incidentally Tornier told me [that] as a "foreign scholar" [the]
institute library etc. are available to me.[532]

Nun herzliche Grüße Ihre Emmy Noether.

Notes

[525]When Noether talks about the "news from Hamburg", she is referring to
information that she had learned 14 days earlier during her stay in Hamburg,
maybe from the Artins with whom she lived, as she reports later in this
letter. Apparently word had gotten to Hamburg that extreme Nazi followers
had prevented Hasse from assuming his new post in Göttingen on his first
attempt. The matter took place on May 29, 1933 and Hasse then returned
to Marburg to wait for the decision of the ministry. Only on July 2nd was
Hasse able to take up his position as institute director in Göttingen. (The
above dates are taken from the book by S.L. Segal [Seg03].)

[526]When Noether says that she cannot invite Hasse now, that reflects the
atmosphere in Göttingen. Hasse was accused several times in the past of
maintaining "contacts with Jews". In the eyes of his opponents that made
him unsuitable for taking over the Göttingen chair. Apparently Noether knew
about it and she wanted to avoid additional problems for Hasse (and maybe
for herself too?) from the extreme Göttingen Nazis. For the same reason she
writes further down in the letter that she probably shouldn't visit Hasse in
Marburg either. Instead of this she suggests a meeting in the house of Hasse's
parents in Bad Soden. The letters do not reveal whether Hasse actually met
with Noether in Bad Soden. Surely they met in July in Göttingen after Hasse
took up his position.

[527]Apparently in the meantime Noether had given up hope of being allowed to return to her beloved Göttingen.

[528]Here S denotes the Frobenius substitution of a prime divisor of $\mathfrak{p}$ and $\langle S \rangle$ the sum of its conjugates in the group ring.

[529]All this concerns Noether's (unsuccessful) attempt to generalize class field theory to non-abelian galois extensions. There is one letter from Artin to Hasse dated August 19, 1927 regarding this topic, but it is not addressed any more in the subsequent Artin-Hasse correspondence.

[530]Witt later, in 1936, provided a proof of the functional equation of the L-series for arbitrary function fields over a finite constant field. And *without* using the Riemann hypothesis, but with the help of a generalization of the Riemann-Roch theorem. However, Witt's proof has not been published. See the collected works of Witt [Wit98], as well as [Roq01].

[531]Witt's paper on this was published in Mathematische Zeitschrift [Wit34b].

[532]If Noether reports that Tornier treated her as a "foreign scholar", then it expresses disappointment and sadness because she still considered Göttingen as her mathematical home. In the meantime, Tornier was installed as Ordinarius and co-director of the Göttingen Mathematics Institute, contrary to the wishes of Hasse, who had other plans for filling this chair. After the Nazis took power in January 1933, Tornier presented himself as a zealous Nazi, which is probably why he got the chair in Göttingen. Over the course of the following years he repeatedly acted against Hasse. In 1936 he was recalled from Göttingen to Berlin after Hasse had threatened to resign from Göttingen because of the defamations by Tornier. – The attitude of Tornier to Emmy Noether is illuminated by the following incident: After the death of Emmy Noether, Hasse, through Hermann Weyl, arranged for a wreath to be placed on her grave as a last salute from her Göttingen colleagues. There is a letter from Tornier to Hasse dated June 27, 1935, in which he agreed to giving his share of the cost of this wreath, but explicitly emphasized that he would not have contributed if he would have known about it beforehand.

3.77   Letter of July 15, 1934

Deuring's habilitation thesis

Göttingen, July 15, 1934

Lieber Herr Hasse!

Deuring sent me a carbon copy of his habilitation thesis to read; I wanted to ask you if I should bring a carbon copy to Gebhardt's Hotel for you, or whether you maybe will come to my place on one of the following evenings? [533]

It is in his interest for Deuring to habilitate as soon as a final decision has been made about his position, for the following reason: In Saxony[534] since the end of the winter semester the provision was made that the state examination must be taken before habilitation could be filed.[535] He's preparing for that now (pedagogics and philosophy), but can only register in the next semester since the required Saxony study semesters are missing: before 1935 nothing [works] with the habilitation! This deadline has just recently been extended by a full year; he must first serve a traineeship at a school before he is admitted to habilitation. So before 1936 there is nothing to hope for.

Now the state examination will soon be required here (as Tornier recently told me), I don't know whether the trainee year [will also be extended]. It is in his interest for Deuring, if possible, to submit an application before this decree is announced; if necessary you can make corrections afterwards.

He is a compulsory member of the dozent academy, now established also in Leipzig, and has reported for camp duty there, in addition, he already has done 3 weeks camp service during the Easter holidays, military sport, in the Erzgebirge. So these prerequisites are met; as far as I know, the camp service and the dozent academy can be completed after the scientific part.[536]

The habilitation thesis is the validity of the Riemann conjecture for imaginary quadratic fields in strips that grow with increasing D.[537]

Herzliche Grüße, Ihre Emmy Noether.

I'm always reachable by phone around $\frac{1}{2}$ 4 p.m. afternoons at 3060, lunchtable Schweiger.

Notes

[533]Hasse had taken up his position in Göttingen on July 2nd, 1934. Apparently he lived weekdays in Gebhardt's Hotel (the one near the train station) and drove to his family in Marburg on the weekend. The date of this letter, that is, July 15, 1934, fell on a Sunday, so Hasse was probably not in Göttingen and therefore Noether had to write a letter.

[534]Deuring had an assistantship with Van der Waerden in Leipzig, i.e. in Saxony. However, for the reasons given by Noether, he was advised not to habilitate in Leipzig but in Gottingen.

[535]Formerly, the state examination was required for those students only who wished to become a teacher at *Gymnasium* (high school). It was controlled by a representative of the state. The habilitation procedure was an examination under the exclusive control of the university faculty. That now a state examination was also required for a habilitation, reflects the fact that the Nazi state authorities wished to assume full control of the university faculties too.

[536]These duties illuminate the methods used by the Nazi regime to check on the political attitudes of prospective *Privatdocent* before admission to the habilitation. – Deuring's habilitation procedure in Göttingen was delayed on formal grounds until December 11, 1935. His scientific standing was acknowledged by the faculty and he received the formal title of "*Privatdozent*" but, for political reasons, he did not get permission to give lectures at the university despite strong support by Hasse and Van der Waerden. This was due to the opposition of some zealous Nazis on the faculty committee, among them the mathematicians Tornier and Teichmüller.

[537]Deuring's habilitation thesis deals with the zeta function of a definite quadratic form Q; let $-D$ denote the discriminant of Q. Deuring shows that there is a constant c such that all zeros of $\zeta_Q(s)$ with imaginary part $0 < t < D^c$ have real part $\frac{1}{2}$. This is what Noether means when she writes "the validity of the Riemann conjecture for imaginary quadratic fields in strips that grow with increasing D". Hasse accepted Deuring's paper for publication in Crelle's Journal; it appeared in 1935 [Deu35c].

3.78   Letter of October 31, 1934

Literature by Lefschetz. Seminar on class field theory at Princeton. Approach to simplification. Artin-Iyanaga's proof of the principal ideal theorem. Taussky. Pyrmont. Weyl. Courant. Ore.

Bryn Mawr College, October 31, 1934

Lieber Herr Hasse,

I got literature from Lefschetz, some of his recent works related to your questions:[538]

1. Report of the Committee on Rational Transformations, published by the National Research Council, Ch. 17

2. S. Lefschetz, L'analyse Situs, last chapter (Collection Borel)

3. S. Lefschetz, Transactions Am. Math. Soc. vol 22, 1921, part 2 (Memoire Bordin)

Perhaps you can give the literature to Schilling,[539] who asked me about algebraic geometry (more variables); apart from Lefschetz there is only Zariski in this direction in U.S.A.

There are a number of interested people in Princeton this year; I'm doing a seminar on class field theory, that is essentially a lecture but with the occasional participation of the audience. We are stuck for the time being, however, still on galois theory; but next time Miss Taussky, who attends from time to time, will present simplest number-theoretic examples. She held a dress rehearsal here; I'm doing the same here tailored to women, i.e. through an incredible diligence on my part replacing the lack of independence in the girls – this year, apart from Miss Taussky, there are two others with scholarships.[540]

Upon reflection, I think that the shortest way, which also requires the fewest assumptions, is a combination of Chevalley with some transcendental methods: One makes the class division according to the (composite) Frobenius-Artin symbol; then to show that the whole group is exhausted, it suffices, for the abelian case, to prove that for cyclic fields of prime degree non-split prime ideals exist. I will do this first transcendentally with the density of the fully split [primes] (thus saving all the counting $h \geq n$ that I might bring after). Then it continues according to Chevalley: Proof that the mapped group [is] equal to the Artin group for cyclotomic fields, and reduction of the general case to it. That gives one at the same time preparation for the hypercomplex formulation: every step can be translated.[541]

I have something more fundamental to think about, occasioned by a talk [I gave] at the New York meeting (where I had 200 people from colleges in and around New York) about modern methods: From the Artin-Iyanaga proof of the principal ideal theorem delete the entire §3, with the exception of 3 lines on the last page.[542] It is shown there: from factor system one it follows that $\overline{G}'$ equals $\overline{U}^{\{...S-1...\}}$, where S runs through all the elements of Γ (because the S and the $\overline{U}$ are interchangeable, so this gives $S^{-1}\overline{U}S\overline{U}^{-1}$ mod commutators). But that means: $(\overline{G}')^{\Gamma} = \overline{U}^{\{...S-1...\}\Gamma} = E$, as was to be shown.[543] This shows even more clearly that the decomposition field alone does everything: I now simply compare the proof method with e.g. the proof of the non-vanishing of the discriminant in algebras without radical. By virtue of the decomposition field Λ one goes to the full matrix ring via Λ (or the direct sum), and reads off everything from the normal basis of the units. Here you read it off from the normal form of factor systems and $\overline{G'}$. I will also send it to Artin; perhaps he can print these three lines.

How was Helsinki?[544] I have heard about other things, but not about this. Naturally I have told a lot about you here; there is a lot of interest, also for Pyrmont.[545] Of course, Weyl is especially interested; but also Courant, very comfortably located in N.Y. – he lives 40 minutes away by car, rural and not far from the beach – becomes a "mensch".[546] Today Rademacher is expected in Philadelphia; after temporary financial difficulties, the situation makes things possible now.[547]

I am very pleased that Ore wants to visit you in January; I hope he draws others after him!

Nun herzliche Grüße Ihre Emmy Noether

Notes

[538]Apparently Hasse had requested references to literature on the foundation of classical algebraic geometry, with a view to his project of proving the Riemann conjecture for function fields of any genus. It seems of interest that already in 1934 Hasse and Lefschetz had contact in this matter, at least indirectly through Emmy Noether. Two years later, at the International Conference in Oslo, they met again and discussed the structure of the endomorphism rings of abelian varieties; see [Roq18].

[539]For Schilling see also the letter from Hasse to Noether dated November 19, 1934.

[540]See Olga Taussky-Todd's Remembrances of Emmy Noether [TT81].

[541]Here Noether seems to outline her lectures in the seminar at Princeton which she mentioned earlier. But in her postcard dated November 23, 1934 Noether admits that the development of class field theory as outlined here probably cannot be carried out in this form.

[542]This is the principal ideal theorem of class field theory. The first proof, due to Furtwängler, [Fur29] was subsequently simplified several times, first by Magnus [Mag34], then by Iyanaga [Iya34]. Noether refers to the latter paper

and uses the same notations. That she calls this the "Artin-Iyanaga-proof" is justified, because in the introduction to his paper, Iyanaga emphasizes that *"The greater part of the paper is due to Artin"*.

[543]Entry from Hasse's hand: *"N.B.* $\overline{G}'^{\Gamma} = E$ is not enough!!! It has to be shown that $\overline{U}^{\Gamma} = E$. But from $\overline{G}'$ it only follows that $\overline{G}' \geq U$, not $\overline{G}' = \overline{U}!$" See also the letter from Hasse to Noether dated November 19, 1934, as well as the two postcards Noether wrote dated November 23 and 28,1934.

[544]See the letter of April 26, 1934.

[545]The annual meeting of the German Mathematical Society (DMV) took place in Bad Pyrmont in September, 1934. What was special about this conference were the political issues. First it was about the so-called *"Gleichschaltung"* of the DMV in the sense of Nazi politics, i.e., the full control of the DMV by Nazi functionaries. This could finally be averted through the intervention by Blascke, Hasse and Knopp, but only to a certain degree. The second issue was the position of the DMV on the controversy between Bieberbach and Harald Bohr. After Bieberbach published his infamous article attacking Harald Bohr in the annual report of the DMV (against the will of the other publishers Hasse and Knopp), this had become a topic for the DMV. Hasse authored a report about the course and the results of the membership meeting at that time, and sent a copy to Hermann Weyl among others. Emmy Noether appears to be referring to this in the present letter. – By the way, Weyl replied to Hasse's report on December 1, 1934. Among other things, he wrote: *"I was very grateful to you that you kept me up to date. With factual success one can probably be satisfied under the circumstances, although the wording of the resolution in the matter of Bieberbach–Bohr does not sound very pleasant in our ears here; it will be for the best if it does not become public ... ."* The events at the Bad Pyrmont conference have often been written about. In his book "Mathematicians under the Nazis" [Seg03] Segal writes about the conference: *"This is perhaps the most frequently discussed occurrence among mathematicians under the Nazi hegemony"*.

[546]This latter remark seems to indicate that the relation between Noether and Courant has become more friendly when compared to their somewhat more formal relation in Göttingen.

[547]Hans Rademacher had been Professor in Breslau. In February 1934 he had been fired by the Nazis, although he did not have a Jewish background. Hasse had recommended Rademacher to Davenport, who then had suggested a job for him at the newly founded school of Kurt Hahn in Gordonstown, Scotland. (Hahn, the founder of the renowned Salem school in Germany,

had to flee from the Nazis.) But as we know from the correspondence Hasse-Rademacher, Emmy Noether had recommended him to the University of Pennsylvania in Philadelphia. Obviously she was informed about Hasse's activities to support Rademacher; this would explain why she mentioned his name here.

3.79 Letter from Hasse of November 19, 1934

Schilling's dissertation. H.L. Schmid. Basing class field theory on the method of Käte Hey and Witt. On the proof of the principal ideal theorem. Bieberbach. Blaschke.

Göttingen, November 19, 1934.

Dear Miss Noether!

Thank you very much for your kind letter of October 31st. I will be happy to take a look at the literature from Lefschetz that you mentioned, when I have a little more peace, and above all also inform Schilling about this. I have told Schilling that he could send you his dissertation for consideration for inclusion in the Mathematische Annalen.[548] I have worked together with Schilling in several many-hour-long sessions to help make detailed corrections and bring the paper from a chaotic condition to one approaching completion. I can definitely not accept it in Crelle, the publisher is very clear about this. The dissertation of H. Schmid written at the same time as Schilling, has been accepted by Knopp for the Mathematische Zeitschrift.[549]

Witt and I are of the opinion that once one uses analysis to justify class field theory – and so far no one has been able to do this without analysis[550] – then the cannon should be aimed straight at headquarters, and in this case that is certainly not the theorem of arithmetic progressions or the inequality $h \leq n$ but the norm theorem, the sum theorem for the invariants and the theorem of everywhere split algebras. Consequence: one uses Hey's cannon in Witt's mount, and then shoots backwards to the classical class field theory as was done for your 50th birthday.[551]

Now, I'm sorry, I have to point out that the master made a mistake. Only trivial things follow from $(\overline{G}')^{\Gamma} = 1$. The real assertion is: $\overline{U}^{\Gamma} = 1$. From your result only the trivial part $U^{\Gamma} = 1$ follows because $\overline{G}' \geq U$. But it isn't certain, as you seem to believe, that $\overline{G}' = U$. It seems to me that § 3 by Iyanaga is not expendable at all. It is only there where really the finiteness condition is used. Your "proof" would also be valid for infinite groups U, and there the theorem is seen to be incorrect due to Witt's counterexample.[552]

I haven't been to Helsingfors. Due to the situation here, I couldn't decide to travel away from here for a longer time. If there is foreign exchange available next spring, I will then make up for it. At the moment we again have to fight against Bieberbach in the D.M.V.[553] You will soon receive a circular from Blaschke.[554]

Warm greetings to all acquaintances and good luck for the spread of class field theory and modern algebra among the Yankees and their daughters.

Always your H. Hasse

Notes

[548]O. F. G. Schilling had been a student of Noether in Göttingen. After Noether's dismissal, she had recommended him to Hasse as a doctoral student. His dissertation was published in the *Mathematische Annalen* [Sch35a]. There he treats the problem given to him by Noether: to compare the arithmetic of a number field K with the arithmetic of a simple algebra having K as a maximal commutative subfield. This problem follows up on the three papers by Hasse, Noether and Chevalley which appeared in the memorial volume for Herbrand [Has34d], [Noe34], [Che36]. – Since Hasse writes that

Schilling should send his dissertation to Noether for the *Mathematische Annalen* we conclude that she still works as an (inofficial) editor of that journal, as she had done for many years in Göttingen.

[549]This is discussing the dissertation of Hermann Ludwig Schmid [Sch35b]. It contains explicit formulas for the norm residue symbol of cyclic extensions of degree p for algebraic function fields of characteristic p.

[550]One year later Chevalley succeeded in building class field theory without analysis [Che35].

[551]Hasse refers to the dissertation by Käte Hey in Hamburg in 1927 with Artin [Hey29]. This is where the zeta function of a division algebra over an algebraic number field is defined and studied, in particular the functional equation. Hey's dissertation contained some errors that were later removed by Max Zorn (a doctoral student of Artin) in a note in the *Hamburger Abhandlungen* [Zor33]. As Zorn pointed out, Hey's results can be used to prove the local-global principle for algebras over number fields. Deuring writes in his book on algebras [Deu35a] that this proof *"represents, as it were, the strongest concentration of analytical tools to achieve the goal"*. After Zorn's note became known Emmy Noether assigned her doctoral student Ernst Witt the task of applying Hey's theory to function fields with finite constant fields. Witt carried this out in his dissertation [Wit34a], at the same time he formulated and proved the Riemann-Roch Theorem for algebras over function fields. Hasse's remark about "Hey's cannon in Witt's mount" is probably to be understood as meaning that Witt's very elegant methods should in turn be transferred to the number field case. The phrase *"shoots backwards to the classical class field theory like was done for your 50th birthday"* refers to Hasse's paper [Has33b] that he had dedicated to Emmy Noether on her 50th birthday in March 1932. There Hasse had shown how Artin's reciprocity law can be derived with the help of the local-global-principle for algebras. – Zorn's results had been mentioned already in earlier letters of Noether in 1931 and 1933.

[552]In the previous letter of October 31, 1934 Noether falsely claimed that the Artin-Iyanaga proof of the principal ideal theorem can still be essentially simplified even further. Hasse now shows that Noether's argument is invalid. The notation is based on Iyanaga [Iya34]. – Witt's counter-example was published in the *Hamburger Abhandlungen* [Wit36a].

[553]The problems of the DMV with Bieberbach in political terms were partially addressed in the notes to Noether's previous letter of October 31, 1934. The "fight" with Bieberbach lasted until February 1935. A letter from Hasse to Hermann Weyl dated February 11, 1935 reads in part [translated]: *"Today I can report to you the following: 2 weeks ago in the German Mathematical Society peace was finally achieved. At a committee meeting in Berlin, Blaschke and Bieberbach resigned at the same time ..."*

[554]This suggests that Emmy Noether had not resigned from the DMV and was therefore listed as a recipient for the circular.

3.80 Postcard of November 23, 1934

Simplification of class field theory won't work. Riemann matrices and endomorphismen ring. Rademacher. Deuring finished his report on algebras.

November 23, 1934

Lieber Herr Hasse!

You are of course quite right that my "simplification" of class field theory does not work. Something, e.g. hypercomplex functional equation, is needed if you throw out the count. On the other hand, my simplification of the principal ideal theorem really is correct! [555]

Have you looked at the? Lefschetz literature?[556] After reading the paper of Weyl "On Riemann Matrices" (in the last issue of the Annals) I am again doubtful if they coincide with your meromorphism rings. That must be decidable![557]

Rademacher has been in Philadelphia since the beginning of November.[558] Deuring writes that his hypercomplex report was sent to Neugebauer at the end of October.[559]

Beste Grüße, Ihre Emmy Noether.

Notes

[555]See the following postcard of November 28, 1934, in which Noether withdraws this assertion after Hasse pointed out her error in his letter of November 19, 1934.

[556]See Noether's letter of October 23, 1934, as well as Hasse's reply on November 19, 1934.

[557]The "Riemann matrices" reflect the period relations for Jacobian manifolds. The corresponding algebras are essentially the endomorphism rings of the Jacobian manifolds, i.e. the "meromorphism rings" in Hasse's terminology. However, Hasse wants to capture these meromorphism rings *algebraically*, and not use the Riemann matrices in their classical definition at all, because by their very nature, periods are *transcendental* constructs. It is not clear whether Noether recognized this connection.

[558]Concerning Rademacher see the previous letter of October 31, 1934.

[559]The correspondence between Hasse and Noether and also other sources indicate that Noether took an avid interest in the creation of Deuring's report [Deu35a] and often advised him in its preparation. We quote Deuring himself, who on June 26, 1934 wrote to Neugebauer as the editor of the series *"Ergebnisse der Mathematik"* that the submission of the manuscript would unfortunately be delayed because *"the largest part of the manuscript is still in the hands of Prof. E. Noether, and I do not want to do without her valuable comments."* It is therefore somewhat surprising that in the foreword to his report Deuring does not mention the contribution of Emmy Noether and her "valuable comments" at all. We can only imagine that Emmy Noether herself advised against using her name publicly, so that Deuring, whom she considered to be her best student and with whom she maintained warm friendly contact, would suffer no unpleasant consequences. In fact, in regards to his habilitation in Göttingen (1935), Deuring was attacked by members of the committee "because of his interaction with Jews". Nevertheless, Hasse could achieve that Deuring was habilitated, but Deuring did not receive from the committee the permission to teach as a *Privatdozent* in Göttingen, despite the apparent wishes of Noether and Hasse.

3.81 Postcard of November 28, 1934

Back to the principal ideal theorem. Class field theory in the USA. Ulm's work on abelian groups.

November 28, 1934

L.H.H.!

I wanted to write to you because I finally woke up – with my postcard the other day I wasn't yet – and saw that I took the beginning of the principal ideal proof for the end, when your letter came. Thank you![1]

The class field theory continues to spread nicely; e.g. Ward, from Pasadena in California, this year at the Institute in Princeton, spends most of his time studying your draft.[2] He along with Zariski are among the professors in the audience; and the latter begins to plunge into the arithmetic theory of algebraic functions! I will gladly present the hypercomplex construction another year – I could not presuppose the discriminant theorem nor anything else – if the listeners have not changed completely. Please ask Ulm again to send his abelian groups; I already have written to him.[3]

Herzliche Grüße, Ihre Emmy Noether.

---

[1]Cf. Noether's letters from October 31 and November 23, 1934 and Hasse's answer dated November 19, 1934.

[2]This refers again to the elaboration of Hasse's Marburg Lectures on class field theory [Has33d].

[3]Helmut Ulm had studied in Göttingen and Bonn and completed his doctorate in Bonn under the direction of Otto Toeplitz in 1933. His two papers on infinite abelian groups became widely known: [Ulm33], [Ulm35]. Probably Noether knew Ulm from his years in Göttingen.

3.82   Letter from Hasse of December 17, 1934

Reply to a greeting card. The Chevalley-Nehrkorn manuscript. Arithmetization of class field theory.

Göttingen, December 17, 1934

Liebe Fräulein Noether,

Thank you very much for your informative letter! I can well imagine that in the stormy weather and useless classes you spend your time dancing on tiptoe, that you are spiritually broken and have no joy in the uncooked meals but are finally happy to be there. That you need me and therefore are willing to sacrifice money, love, beer, sleep, food and holidays is a great honor for me. Whereas I am ashamed that you have nothing to thank me for.[560]

Now seriously: At the request of Mr. Nehrkorn I am enclosing the manuscript of his joint work with Chevalley for the *Mathematische Annalen*.[561] From the standpoint of a purely arithmetical foundation of class field theory it seems to me that the maximum has been achieved. I don't think that without a whole new method a purely arithmetical proof of the main theorem of algebras will succeed.[562]

With best wishes for a Merry X-mas and a Happy New Year

Ihr H. Hasse

P.S. Please give my greetings also to the signatories of your content-rich letter. I would be very grateful to you for the transmission of a number of blank checks of this kind to recompense my valuable time.[563]

Notes

[560]Apparently this is the reply to a humorous greeting card (which has not survived) of the kind that might be sent from a party. We can implicitly read the content of Noether's remarks from Hasse's reply. We are reminded of the report from Olga Taussky about her contact with Emmy Noether during these months: *"I do not wish to give the impression that she was in a bad or depressed mood all that academic year. But certainly she was in a very changeable mood."* And the report from Brauer to Hasse dated April 18, 1935: *"The only cloudiness for Emmy was that strong homesickness for Göttingen and her friends... she spoke very much about you."*

[561]Chevalley and Nehrkorn were both in the audience at Artin's lecture in Göttingen in the year 1932. In their paper they showed that class field theory can be derived purely arithmetically (i.e. without analytical tools) from the summation formula for the Hasse invariants of an algebra. The work appeared in *Mathematische Annalen* in the year 1935 [CN35]. The fact that Hasse sent the manuscript to Emmy Noether shows that Emmy Noether still, even after her forced emigration, was regarded as an (unofficial) editor of the *Mathematische Annalen*.

[562]In the same year 1935, however, Chevalley announced in the Comptes Rendus Paris that he had succeeded in developing class field theory completely without analytical tools: [Che35]. Indeed, Chevalley used a completely new idea which he had developed together with his colleage Herbrand (who unfortunately had died in the meantime). See, e.g., [Roq14].

[563]Here, Hasse returns to the joking tone of the first sentences of his letter.

3.83   Letter of April 7, 1935

Letter from H. to Albert. Quadr.forms and algebras. Dirac. Weyl. Brewer. Extension cycl. field to cycl. field. Explicit construct of normal bases. Diss. Stauffer. Helsinki? Summer plans.

Bryn Mawr College, April 7, 1935

Lieber Herr Hasse,

Thank you for your separate mailing – I always admire how much you manage – and for the copy of the letter to Albert.[564]   The additive crossed products and the "summand systems" – some of these things have already appeared in the the latest Crelle paper by Witt – have given me a lot of pleasure.[565] Brauer draws attention to the relation between quadratic forms and algebras (p. 18) that includes a proof of the product formula.[566] $H = \prod_{i \le k}(a_i, a_k)$ (Artin) is well known to physicists: Dirac used this, and Weyl and Brauer used this in a note on the theory of spinors.[567] The special case of the sum of squares is said to be in the article about hypercomplex systems by Cartan in the French encyclopedia.

It is known that the splitting of factor systems in the multipl[icative] case is [a] necessary and sufficient condition for the existence of the extension. Witt likely knows? That's the result by Brauer, which you mention in annual report 44.[568]

Brauer has pretty good prospects to come to Toronto as an assistant professor; the trustees only have to approve it.

The elegant successive construction of the cyclic normal form seems considerably more reasonable to me than a confusing direct [construction]; the successive "summand systems" are just the invariants.[569]

In connection with this, you may be interested in an explicit statement regarding "normal basis", which in the meantime I had my doctoral student here carry out for the case that the characteristic of K/k does not divide the degree.[570]

Let $E^{(1)}$, ..., $E^{(r)}$ be the irreducible central idempotents of the group ring $(G)_k$ of the galois group of K/k, with k as coefficients. Then for each $E^{(i)}$ there is an element z_i in K, so that $E^{(i)} z_i \ne 0$. (If $E^{(i)} = \sum S_\lambda \alpha_\lambda$, this means: $\sum z_i^{S_\lambda} \alpha_\lambda \ne 0$). Each element $w = \sum_i E^{(i)} z_i$ (no component zero) is the carrier of a normal basis – the w^S linear ind[ependent] – and so you get *all* normal bases. In particular, starting from some basis z_1, ..., z_n, you can choose the z_i [from above] from these; thus giving an explicit rational construction.

In the case that p divides degree, one will presumably have to go back to the composition series for the simplest representation, similar to Witt in the cyclic case.

Another result is the determination of discriminant of the center of the integral group ring – i.e. the classes of conj[ugate] elements as basis – to $n^r \cdot \prod_1^r h_i \big/ \prod_1^r f_i^2$. where r means the rank of the center (number of ab[solutely] irreducible representations), h_i the number of elements in the individual classes of conjugate elements, f_i the degrees of ab[solutely] irreducible representations. In particular, then, this fraction is an integer, but this may also follow from known formulas (it is weaker than [requiring] that the degrees f_i divide n; i.e. neither of the two results seems to follow from the other). I have derived this and known formulas from the theory of complementary bases.[571]

Have you been to Helsinki[572] And what do you do in the summer? I do not yet know whether I will come back this year; if so it will [be] end [of] June. I must attend the "commencement" (graduation ceremony) and ceremoniously present my doctoral student for the doctoral hat (a kind of hood); she gets the hood placed on her by the President.

Herzliche Grüße, Ihre Emmy Noether.[573]

Notes

[564]A copy of this letter from Hasse to Albert, dated February 2, 1935, is in the Hasse estate. Its 21 pages contain a detailed report on recent results from the circle around Hasse in Göttingen, among them the presentation of a preliminary step in the theory of Witt vectors for the construction of cyclic extensions of p-power degree over a base field of characteristic p.

[565]A "summand system" is the additive analog of a multiplicative "factor system" in the sense of Emmy Noether. In today's terminology, these are 2-cycles in an additive group. Witt had proved that for a Galois extension $K|k$ each summand system of the galois group splits, that is, the 2-cohomology of the additive group K^+ vanishes. This theorem can be found in [Wit35a]. We observe that Witt's paper was published in Crelle's Journal in the year 1935. It appears improbable that this volume had already been available in the USA at the time of Noether's letter. Hence we conclude that Witt (or Hasse) had informed Noether about Witt's results earlier already. Hasse

had used these "summand systems" in his paper [Has34c] in which he gave a foundation of class field theory for global fields of characteristic $p > 0$, including the arithmetic of Artin-Schreier extensions in characteristic $p > 0$. It appears that Hasse had sent Noether a reprint of this paper.

[566]In the letter from Hasse to Albert mentioned above, to which Noether refers, the following text can be found on pages 16–18:

"*You will remember my writing you about a certain connection between general quadratic forms and linear algebra. Two years ago Artin proved that for a quadratic form*

$$f(x) \;=\; a_1 x_1^2 + \cdots + a_n x_n^2 , \quad a_i \neq 0$$

with coefficients a_i in an arbitrary field k (not of characteristic 2) the normal simple algebra of exponent 2 and order $4^{\frac{n(n+1)}{2}}$ over k:

$$H \;=\; \prod_{i \leq k} (a_i, a_k)$$

is an invariant. Here (a, b) denotes the generalized quaternion algebra over k generated by

$$u^2 = a , \quad v^2 = b , \quad vu = -uv .$$

Artin's proof depends on certain identities between minors of a general symmetric matrix $A = (a_{ik})$. Witt found a very much nicer and simpler proof, that, moreover, explicitly shows the close connection between the form $f(x)$ and the algebra H . . . ."

Hasse then gives a sketch of Witt's proof. Today H is called the "Hasse algebra" of the quadratic form f since in the case of a local number field it essentially goes back to Hasse's early paper on the equivalence of quadratic forms. Artin had given his proof in a lecture, but apparently never published it. Witt's proof was later published in [Wit37]. We see that Emmy Noether knew also about an early version of this paper of Witt's.

[567]This is the paper [BW35] entitled: "Spinors in n dimensions". At that time Brauer worked at the Institute for Advanced Study in Princeton as assistant to Hermann Weyl.

[568]This concerns the question of whether a cyclic extension $k|k_0$ of prime power degree p^{h-1} can be embedded into a cyclic extension $K|k_0$ of degree p^h. Hasse reports in his letter to Albert that Witt can answer this question: In the case of the characteristic p this is always possible, however, in the case of the characteristic $\neq p$ not always, and there will be a necessary and sufficient condition. If k_0 contains a primitive p-th root of unity ζ, this condition is that ζ is a norm from k.

This can be interpreted as the splitting of the factor system of $k|k_0$ defined by ζ. Noether now points out that this splitting condition is known from [Bra32b], and asks if Witt knows that. She refers to Hasse's solution of a problem which had been posed by Van der Waerden, namely: *"Which quadratic number fields can be embedded in a cyclic field of degree 4?"* In the annual report of the DMV (volume 44, 1934) Hasse gives three solutions to this problem: one based on class field theory, one based on Kummer theory, and one using the theory of algebras. Following the algebra-theoretic solution, Hasse (relying on a communication from Van der Waerden) gives a solution to the general problem of embedding cyclic fields in cyclic fields, with reference to the theorem quoted by Noether in [Bra32b]. – By the way, in the same volume of the annual report, Hasse sets and solves problem 169: *"Under which conditions can a cyclic field K_1 of prime degree ℓ over an algebraic number field K_0 be embedded in a cyclic field K_n over K_0 of degree ℓ^n?"* The receipt date for this problem is October 10, 1933, which is one week earlier than the receipt date October 17, 1933 of Hasse's three solutions to Van der Waerden's problem mentioned above.

[569] As Hasse reported in his letter to Albert, Witt succeeded in constructing a cyclic extension of degree p^h by successive steps of degree p. To each such step there is a "summand system" in the smaller field that splits in the larger field lying above it. These summand systems are connected to each other by virtue of the trace operator and the Artin-Schreier operator. That's what Noether apparently means when she talks about the "successive construction".

A year later, in the summer semester, 1936, Hasse's legendary "working group" took place in Göttingen, in which Witt found, among other things, the formalism known today as *Witt vectors* [Wit36b]. Based on Witt vectors, in the cited paper Witt actually gave a "direct" and at the same time particularly "elegant" normal form for cyclic p extensions. If Noether had been able to take note of that, then she certainly would have preferred the direct normal form with Witt vectors to the "successive" normal form.

[570] The doctoral student was Ruth Stauffer. Her dissertation (Bryn Mawr) appeared in the American Journal of Mathematics [Sta36]. – The existence of a normal basis for an arbitrary galois field extension was probably first proved by Emmy Noether [Noe32b]; also see the letter dated August 22, 1931 and the following letters. There the base field was assumed to be infinite; later Deuring [Deu32] gave a proof without this assumption. However, it was just an existence proof. The Stauffer dissertation is about the establishment of an explicit algorithm for constructing a normal basis. In the paragraph following this footnote, Noether gives a sketch of Stauffer's construction.

[571] For this formula, E. Dade said the following when asked: *"It is straightforward and would surely have been possible at the time of Burnside and*

Frobenius if any one had just asked the question. Apparently nobody asked the question about the discriminant at the time."

[572] As mentioned in a note to letter of April 26, 1934, Hasse had planned to accept an invitation to give lectures in the fall of 1934 in Helsinki. However, he had to cancel this trip. On August 13, 1934 Hasse had written to Nevanlinna:

> *"The past few months have caused a lot of excitement related to my appointment to Göttingen . I had hoped that by September I would be over all this, both internally and externally. But I see now that that won't be the case... After careful consideration, I have come to the conclusion that it is better if I move my intended trip to you to a later time... I don't feel able to immerse myself in scientific questions... And finally, at the moment I don't feel inwardly up to the task of appearing abroad as representative of German science ..."*

The turbulent conditions in Göttingen and in the German Mathematicians Association (DMV) after the Nazi government took power have often been described and therefore need not be repeated here. Hasse finally made the trip to Finland in October 1938.

[573] This is the last letter from Emmy Noether to Hasse. She died one week later. Noether's letter shows that she kept her mathematical interest and activity until the very end.

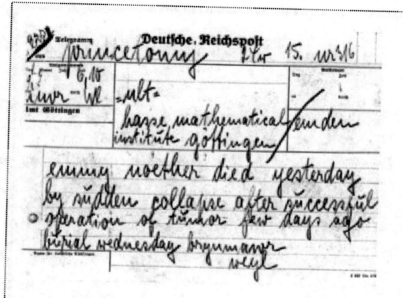

Copy of telegram message from Hermann Weyl (Princeton) to Helmut Hasse (Göttingen) containing death announcement of Emmy Noether who had passed away on April 14, 1935.

Bibliography

[AH32] A.A. Albert, H. Hasse, A determination of all normal division algebras over an algebraic number field. Trans. Am. Math. Soc. **34**, 722–726 (1932)

[AH35] P. Aleksandroff, H. Hopf, *Topologie. Bd. 1. Grundbegriffe der mengentheoretischen Topologie. Topologie der Komplexe. Topologische Invarianzsätze und anschliessende Begriffsbildungen. Verschlingung im n–dimensionalen euklidischen Raum. Stetige Abbildungen von Polyedern.* Die Grundlehren d. math. Wiss. in Einzeldarstell. mit besonderer Berücksichtigung d. Anwendungsgebiete, vol. 45 (Springer, Berlin, 1935), XIII, 636 pp., 39 Abb

[Alb30] A.A. Albert, New results in the theory of normal division algebras. Trans. Am. Math. Soc. **32**, 171–195 (1930)

[Alb31a] A.A. Albert. On direct products. Trans. Am. Math. Soc. **33**, 690–711 (1931)

[Alb31b] A.A. Albert, On direct products, cyclic division algebras, and pure Riemann matrices. Trans. Am. Math. Soc. **33**, 219–234, Correction p. 999, 1931. Remark: This paper has not been included into the "Collected Papers" of A. A. Albert

© Springer Nature Switzerland AG 2022
P. Roquette, F. Lemmermeyer, *The Hasse - Noether Correspondence 1925 - 1935*, History of Mathematics Subseries 2317,
https://doi.org/10.1007/978-3-031-12880-6

[Alb31c] A.A. Albert, The structure of pure Riemann matrices with non-commutative multiplication algebras. Rend. Circ. Mat. Palermo, II.Ser. **55**, 57–115 (1931)

[Alb32a] A.A. Albert, On the construction of cyclic algebras with a given exponent. Am. J. Math. **54**, 1–13 (1932)

[Alb32b] A.A. Albert, On the construction of cyclic algebras with a given exponent. Am. J. Math. **54**, 1–13 (1932)

[Alb34] A.A. Albert, On the construction of Riemann matrices. I. Ann. Math. (2) **35**, 1–28 (1934)

[Arc28] R.G. Archibald, Diophantine equations in division algebras. Trans. Am. Math. Soc. **30**, 819–837 (1928)

[Art23] E. Artin, Über eine neue Art von L-Reihen. Abh. Math. Semin. Univ. Hamb. **3**, 89–108 (1923)

[Art24a] E. Artin, Quadratische Körper im Gebiete der höheren Kongruenzen I. (Arithmetischer Teil.). Math. Zeitschr. **19**, 153–206 (1924)

[Art24b] E. Artin, Quadratische Körper im Gebiete der höheren Kongruenzen II. (Analytischer Teil.). Math. Z. **19**, 207–246 (1924)

[Art27] E. Artin, Beweis des allgemeinen Reziprozitätsgesetzes. Abh. Math. Semin. Univ. Hamb. **5**, 353–363 (1927)

[Art28a] E. Artin, Zur Arithmetik hyperkomplexer Zahlen. Abh. Math. Semin. Univ. Hamb. **5**, 261–289 (1928)

[Art28b] E. Artin, Zur Theorie der hyperkomplexen Zahlen. Abh. Math. Semin. Univ. Hamb. **5**, 251–260 (1928)

[Art29] E. Artin, Idealklassen in Oberkörpern und allgemeines Reziprozitätsgesetz. Abh. Math. Semin. Univ. Hamb. **7**, 46–51 (1929)

[Art30] E. Artin, Zur Theorie der *L*-Reihen mit allgemeinen Gruppencharakteren. Abh. Math. Semin. Univ. Hamb. **8**, 292–306 (1930). Collected Papers, 1965, pp. 165–179.

[Art31] E. Artin, Die gruppentheoretische Struktur der Diskriminanten algebraischer Zahlkörper. J. Reine Angew. Math. **164**, 1–11 (1931)

[AT68] E. Artin, J. Tate, *Class Field Theory* (W. A. Benjamin, Inc., New York, Amsterdam, 1968), 259 pp. Reprint of the notes of the Artin-Tate seminar on class field theory at Princeton University 1951–1952

[AvdW26] E. Artin, B.L. van der Waerden, Die Erhaltung der Kettensätze der Idealtheorie bei beliebigen endlichen Körpererweiterungen. Nachrichten Göttingen **1926**, 23–27 (1926)

[AW45] E. Artin, G. Whaples, Axiomatic characterization of fields by the product formula for valuations. Bull. Am. Math. Soc. **51**, 469–492 (1945)

[BHN32] R. Brauer, H. Hasse, E. Noether, Beweis eines Hauptsatzes in der Theorie der Algebren. J. Reine Angew. Math. **167**, 399–404 (1932)

[BN27] R. Brauer, E. Noether, Über minimale Zerfällungskörper irreduzibler Darstellungen. Sitzungsberichte Akad. Berlin **1927**, 221–228 (1927)

[Bra26] R. Brauer, Über Zusammenhänge zwischen arithmetischen und invariantentheoretischen Eigenschaften von Gruppen linearer Substitutionen. Sitzungsber. Preuss. Akad. Wiss. **1926**, 410–416 (1926)

[Bra28a] H. Brandt, Idealtheorie in einer Dedekindschen Algebra. Jahresber. Dtsch. Math.-Ver. **37** 2.Abt., 5–7 (1928)

[Bra28b] R. Brauer, Untersuchungen über die arithmetischen Eigenschaften von Gruppen linearer Substitutionen. Math. Z. **28**, 677–696 (1928)

[Bra29] R. Brauer, Über Systeme hyperkomplexer Zahlen. Math. Z. **30**, 79–107 (1929)

[Bra30] R. Brauer, Untersuchungen über die arithmetischen Eigenschaften von Gruppen linearer Substitutionen. II. Math. Z. **31**, 733–747 (1930)

[Bra32a] R. Brauer, Über die algebraische Struktur von Schiefkörpern. J. Reine Angew. Math. **166**, 241–252 (1932)

[Bra32b] R. Brauer, Über die Konstruktion der Schiefkörper, die von endlichem Rang in bezug auf ein gegebenes Zentrum sind. J. Reine Angew. Math. **168**, 44–64 (1932)

[Bra33] R. Brauer, Über den Index und den Exponenten von Divisionsalgebren. Tôhoku Math. J. **37**, 77–87 (1933)

[Bra47] R. Brauer, Applications of induced characters. Am. J. Math. **69**, 709–716 (1947)

[BW35] R. Brauer, H. Weyl, Spinors in n dimensions. Am. J. Math. **1935**, 425–449 (1935)

[CF67] J.W.S. Cassels, A. Fröhlich, *Algebraic Number Theory* (Academic, London, New York, 1967), XV, 366 pp.

[Che26] N. (Tschebotareff) Chebotarev, Die Bestimmung der Dichtigkeit einer Menge von Primzahlen, welche zu einer gegebenen Substitutionsklasse gehören. Math. Ann. **95**, 191–228 (1926)

[Che31] C. Chevalley, Sur un théorème de M. Hasse. C. R. Acad. Sci., Paris **191**, 369–370 (1931)

[Che32] C. Chevalley, La structure de la théorie du corps de classes. C. R. Acad. Sci., Paris **194**, 766–769 (1932)

[Che33a] C. Chevalley, La théorie du symbole de restes normiques. J. Reine Angew. Math. **169**, 140–157 (1933)

[Che33b] C. Chevalley, Sur la théorie du corps de classes dans les corps finis et les corps locaux. J. Fac. Sci. Univ. Tokyo, Sect. I **2**, 365–476 (1933)

[Che34] C. Chevalley, Sur certains idéaux d'une algèbre simple. Abh. Math. Semin. Univ. Hamb. **10**, 83–105 (1934)

[Che35] C. Chevalley, Sur la théorie du corps de classes. C. R. Acad. Sci., Paris **201**, 632–634 (1935)

[Che36] C. Chevalley, L'arithmétique dans les algèbres de matrices. (Exposés mathématiques, publiés à la mémoire de Jacques Herbrand, XIV.). Actual. Sci. Ind. **323**, 33 pp. (1936)

[CN35] C. Chevalley, H. Nehrkorn, Sur les démonstrations arithmétiques dans la théorie du corps de classes. Math. Ann. **111**, 364–371 (1935)

[Coh78] H. Cohn, *A classical invitation to algebraic numbers and class fields. With two appendices by Olga Taussky: "Artin's 1932 Göttingen lectures on class field theory" and "Connections between algebraic number theory and integral matrices."*. Universitext (Springer–Verlag, New York, Heidelberg, Berlin, 1978), XIII, 328 pp.

[Cur99] C. Curtis, *Pioneers of representation theory: Frobenius, Burnside, Schur and Brauer*. History of Mathematics (American Mathematical Society, Providence, RI, 1999), XVI, 287 pp.

[Ded95] R. Dedekind, Ueber eine Erweiterung des Symbols $(\mathfrak{a}, \mathfrak{b})$ in der Theorie der Moduln. *Nachrichten Göttingen* (1895), pp. 183–208

[Ded31] R. Dedekind,   *Gesammelte mathematische Werke.*
Zweiter Band. Hrsg. v. Robert Fricke, Emmy Noether
u. Öystein Ore (Friedr. Vieweg & Sohn A.-G., Braunschweig, 1931)

[Ded32] R. Dedekind,   *Gesammelte mathematische Werke.*
Hrsg. v. Robert Fricke, Emmy Noether u. Öystein Ore.
Bd.1–3 (Friedr. Vieweg & Sohn A.-G., Braunschweig,
1930–1932), 397/442/508 pp.

[Deu31a] M. Deuring, Verzweigungstheorie bewerteter Körper.
Math. Ann. **105**, 277–307 (1931)

[Deu31b] M. Deuring, Zur arithmetischen Theorie der algebraischen Funktionen. Math. Ann. **106**, 77–102 (1931)

[Deu31c] M. Deuring, Zur Theorie der Normen relativzyklischer
Körper. *Nachr. Ges. Wiss. Göttingen, Math.—Phys.*
Kl. I (1931), pp. 246–247

[Deu32] M. Deuring, Galoissche Theorie und Darstellungstheorie. Math. Ann. **107**, 140–144 (1932)

[Deu33] M. Deuring, Imaginäre quadratische Zahlkörper mit
der Klassenzahl 1. Math. Z. **37**, 405–415 (1933)

[Deu35a] M. Deuring, *Algebren.* Erg. d. Math. u. ihrer Grenzge-
biete (Julius Springer, Berlin, 1935), 143 pp.

[Deu35b] M. Deuring, Neuer Beweis des Bauerschen Satzes. J.
Reine Angew. Math. **173**, 1–4 (1935)

[Deu35c] M. Deuring, Zetafunktionen quadratischer Formen. J.
Reine Angew. Math. **172**, 226–252 (1935)

[Deu36] M. Deuring,   Anwendungen der Darstellungen von
Gruppen durch lineare Substitutionen auf die galoissche Theorie. Math. Ann. **113**, 40–47 (1936)

[Deu49] M. Deuring, Algebraische Begründung der komplexen
Multiplikation.   Abh. Math. Semin. Univ. Hamb.
16(1/2), 32–47 (1949)

[DH34] H. Davenport, H. Hasse, Die Nullstellen der Kongruenzzetafunktionen in gewissen zyklischen Fällen. J. Reine Angew. Math. **172**, 151–182 (1934)

[Dic23] L.E. Dickson, *Algebras and Their Arithmetics* (University of Chicago Press, Chicago, 1923), XII, 241 pp.

[Dic27] L.E. Dickson, *Algebren und ihre Zahlentheorie. Mit einem Kapitel über Idealtheorie von A. Speiser.* Orell Füssli (Veröffentlichungen der Schweizer Math. Ges. Bd. 4), Zürich, 1927. Übersetzt von J.J. Burckhardt und E. Schubarth

[Dic28] L.E. Dickson, New division algebras. Bull. Am. Math. Soc. **34**, 555–560 (1928)

[Dic70] A. Dick, *Emmy Noether 1882–1935.* Beiheft No. 13 zur Zeitschrift "Elemente der Mathematik" (Birkhäuser-Verlag, Basel, 1970). English translation 1981 by H.I. Blocher

[DW82] R. Dedekind, H. Weber, Theorie der algebraischen Funktionen einer Veränderlichen. J. Reine Angew. Math. **92**, 181–290 (1882)

[Fit32] H. Fitting, Die Theorie der Automorphismenringe Abelscher Gruppen und ihr Analogon bei nicht kommutativen Gruppen. Math. Ann. **107**, 514–542 (1932)

[FL05] M. Fontana, K.A. Loper, An historical overview of Kronecker function rings, Nagata rings, and related star and semistar operations. Preprint, 107 (2005)

[FLR14] G. Frei, F. Lemmermeyer, P. Roquette (eds.) *Emil Artin and Helmut Hasse. Their correspondence 1923–1958. English version, revised and enlarged..* Contributions in Mathematical and Computational Science., vol. 5 (Springer, Basel, 2014), X + 484 pp.

[Fro96] G. Frobenius, Über Beziehungen zwischen den Primidealen eines algebraischen Körpers und den Substitutionen seiner Gruppe. Sitzungsberichte Akad. Berlin **1896**, 689–703 (1896)

[Frö83] A. Fröhlich, *Galois Module Structure of Algebraic Integers* (Springer, Berlin, Heidelberg, 1983)

[FS05] D. Fenster, J. Schwermer, A delicate collaboration: Adrian Albert and Helmut Hasse and the Principal Theorem in Division Algebras in the early 1930's. Arch. History Exact Sci. **59**, 349–379 (2005)

[Fur12] Ph. Furtwängler, Die Reziprozitätsgesetze für Potenzreste mit Primzahlexponenten in algebraischen Zahlkörpern, II. Math. Ann. **72**, 346–386 (1912)

[Fur29] Ph. Furtwängler, Beweis des Hauptidealsatzes für die Klassenkörper algebraischer Zahlkörper. Abh. Math. Semin. Univ. Hamb. **7**, 14–36 (1929)

[Gau00] C.F. Gauss, Zwei Notizen über die Auflösung der Kongruenz $xx + yy + zz \equiv 0 \bmod p$, in *Gauß Werke VIII, Arithmetik und Algebra, Nachträge.*, ed. by G.D. Wiss. Göttingen (Teubner, Leipzig, 1900), pp. 3–4

[Gey66] W.-D. Geyer, Ein algebraischer Beweis des Satzes von Weichold über reelle Funktionenkörper, in *Algebraische Zahlentheorie*, ed. by H. Hasse, P. Roquette. Ber. Tagung math. Forschungsinst. Oberwolfach, vol. 2, Mannheim, 1966 (Bibliographisches Institut.), pp. 83–98

[Gey77] W.-D. Geyer, Reelle algebraische Funktionen mit vorgegebenen Null- und Polstellen. Manuscr. Math. **22**, 87–103 (1977)

[Gre27a] H. Grell, Beziehungen zwischen Idealen verschiedener Ringe. Math. Ann. **97**, 490–523 (1927)

[Gre27b] H. Grell, Zur Theorie der Ordnungen in algebraischen Zahl- und Funktionenkörpern. Math. Ann. **97**, 524–558 (1927)

[Gre30a] H. Grell. Zur Normentheorie in hyperkomplexen Systemen. J. Reine Angew. Math. **162**, 60–62 (1930)

[Gre30b] H. Grell, Zur Verzweigungstheorie in maximalen Ordnungen Dedekindscher hyperkomplexer Systeme und in allgemeinen Ordnungen algebraischer Zahl- und Funktionenkörper. Jahresber. Dtsch. Math.-Ver. **39**(2.Abteilung), 17–18 (1930)

[Gre35] H. Grell, Über die Gültigkeit der gewöhnlichen Idealtheorie in endlichen algebraischen Erweiterungen erster und zweiter Art. Math. Z. **40**, 503–505 (1935)

[Gre36] H. Grell, Verzweigungstheorie in allgemeinen Ordnungen algebraischer Zahlkörper. Math. Z. **40**, 629–657 (1936)

[Gru32] W. Grunwald, Charakterisierung des Normenrestsymbols durch die $\wp$-Stetigkeit, den vorderen Zerlegungssatz und die Produktformel. Math. Ann. **107**, 145–164 (1932)

[Gru33] W. Grunwald, Ein allgemeines Existenztheorem für algebraische Zahlkörper. J. Reine Angew. Math. **169**, 103–107 (1933)

[Has23] H. Hasse, Über die Darstellbarkeit von Zahlen durch quadratische Formen im Körper der rationalen Zahlen. J. Reine Angew. Math. **152**, 129–148 (1923)

[Has24] H. Hasse, Darstellbarkeit von Zahlen durch quadratische Formen in einem beliebigen algebraischen Zahlkörper. J. Reine Angew. Math. **153**, 113–130 (1924)

[Has25] H. Hasse, Zwei Existenztheoreme über algebraische Zahlkörper. Math. Ann. **95**, 229–238 (1925)

[Has26a] H. Hasse,   Bericht über neuere Untersuchungen
         und Probleme aus der Theorie der algebraischen
         Zahlkörper. I: Klassenkörpertheorie. Jahresber. Dtsch.
         Math.-Ver. **35**, 1–55 (1926)

[Has26b] H. Hasse,   Bericht über neuere Untersuchungen
         und Probleme aus der Theorie der algebraischen
         Zahlkörper. I: Klassenkörpertheorie. Jahresber. Dtsch.
         Math.-Ver. **35**, 1–55 (1926)

[Has26c] H. Hasse, *Höhere Algebra. Bd. I: Lineare Gleichungen.*
         Sammlung Göschen, vol. 931 (Walter de Gruyter & Co.,
         Berlin, 1926), 160 pp.

[Has26d] H. Hasse, Neue Begründung der komplexen Multiplika-
         tion I: Einordnung in die allgemeine Klassenkörperthe-
         orie. J. Reine Angew. Math. **157**, 115–139 (1926)

[Has26e] H. Hasse,   Über die Einzigkeit der beiden Funda-
         mentalsätze der elementaren Zahlentheorie. J. Reine
         Angew. Math. **155**, 199–220 (1926)

[Has26f] H. Hasse,   Zwei Existenztheoreme über algebraische
         Zahlkörper. Math. Annalen **95**, 229–238 (1926)

[Has27a] H. Hasse,   Bericht über neuere Untersuchungen
         und Probleme aus der Theorie der algebraischen
         Zahlkörper. Teil Ia: Beweise zu I. Jahresber. Dtsch.
         Math.-Ver. **36**, 233–311 (1927)

[Has27b] H. Hasse, Existenz gewisser algebraischer Zahlkörper.
         Sitzungsberichte Akad. Berlin **1927**, 229–234 (1927)

[Has27c] H. Hasse, *Höhere Algebra. Bd. II: Gleichungen höheren*
         Grades. Sammlung Göschen, vol. 932 (Walter de
         Gruyter & Co., Berlin, 1927), 160 pp.

[Has27d] H. Hasse,   Über das Reziprozitätsgesetz der *m*-ten
         Potenzreste.   J. Reine Angew. Math. **158**, 228–259
         (1927)

[Has28] H. Hasse, Über eindeutige Zerlegung in Primelemente oder in Primhauptideale in Integritätsbereichen. J. Reine Angew. Math. **159**, 3–12 (1928)

[Has30a] H. Hasse, Bericht über neuere Untersuchungen und Probleme aus der Theorie der algebraischen Zahlkörper. Teil II: Reziprozitätsgesetz. Jahresber. Dtsch. Math.-Ver. **6**(Ergänzungsband) (1930), IV + 204 pp.

[Has30b] H. Hasse, Die moderne algebraische Methode. Jahresber. Dtsch. Math. Ver. **31**, 22–34 (1930). Reprinted in English translation in the *Mathematical Intelligencer*, vol. 8 (1986)

[Has30c] H. Hasse, Die Normenresttheorie relativ-Abelscher Zahlkörper als Klassenkörpertheorie im Kleinen. J. Reine Angew. Math. **162**, 145–154 (1930)

[Has30d] H. Hasse, Neue Begründung und Verallgemeinerung der Theorie des Normenrestsymbols. J. Reine Angew. Math. **162**, 134–144 (1930)

[Has31a] H. Hasse, Beweis eines Satzes und Widerlegung einer Vermutung über das allgemeine Normenrestsymbol. Nachr. Ges. Wiss. Göttingen, Math.—Phys. Kl. I (1931), pp. 64–69

[Has31b] H. Hasse, Neue Begründung der komplexen Multiplikation. II. Aufbau ohne Benutzung der allgemeinen Klassenkörpertheorie. J. Reine Angew. Math. **165**, 64–88 (1931)

[Has31c] H. Hasse, Theorie der zyklischen Algebren über einem algebraischen Zahlkörper. Nachr. Ges. Wiss. Göttingen, Math.—Phys. Kl. I (1931), pp. 70–79

[Has31d] H. Hasse, Über $\wp$-adische Schiefkörper und ihre Bedeutung für die Arithmetik hyperkomplexer Zahlsysteme. Math. Ann. **104**, 495–534 (1931)

[Has32a] H. Hasse, Additional note to the author's "Theory of cyclic algebras over an algebraic number field". Trans. Am. Math. Soc. **34**, 727–730 (1932)

[Has32b] H. Hasse, Ansprache zum 70. Geburtstag des Geh. Regierungsrats Prof. Dr. Kurt Hensel am 29. Dezember 1931. Mitteilungen Universitätsbund Marburg **1932**(Heft 1), 1–6 (1932)

[Has32c] H. Hasse, Theory of cyclic algebras over an algebraic number field. Trans. Am. Math. Soc. **34**, 171–214 (1932)

[Has32d] H. Hasse, Zu Hilberts algebraisch–zahlentheoretischen Arbeiten, in *D. Hilbert, Gesammelte Abhandlungen.*, vol. 1 (Springer, Berlin, 1932), pp. 528–535

[Has32e] H. Hasse, Zwei Bemerkungen zu der Arbeit "Zur Arithmetik der Polynome" von U. Wegner. Math. Ann. **106**, 455–456 (1932)

[Has33a] H. Hasse, Beweis des Analogons der Riemannschen Vermutung für die Artinschen und F. K. Schmidtschen Kongruenzzetafunktionen in gewissen elliptischen Fällen. Vorläufige Mitteilung. Nachr. Ges. Wiss. Göttingen, Math.—Phys. Kl. I **1933**(42), 253–262 (1933)

[Has33b] H. Hasse, Die Struktur der R. Brauerschen Algebrenklassengruppe über einem algebraischen Zahlkörper. Insbesondere Begründung der Theorie des Normenrestsymbols und Herleitung des Reziprozitätsgesetzes mit nichtkommutativen Hilfsmitteln. Math. Ann. **107**, 731–760 (1933)

[Has33c] H. Hasse, Explizite Konstruktion zyklischer Klassenkörper. Math. Ann. **109**, 191–195 (1933)

[Has33d] H. Hasse, *Vorlesungen über Klassenkörpertheorie.* Preprint, Marburg. [Later published in book form by Physica Verlag Würzburg (1967)], 1933

[Has34a] H. Hasse, Abstrakte Begründung der komplexen Multiplikation und Riemannsche Vermutung in Funktionenkörpern. Abh. Math. Semin. Univ. Hamb. **10**, 325–348 (1934)

[Has34b] H. Hasse, Normenresttheorie galoisscher Zahlkörper mit Anwendungen auf Führer und Diskriminante abelscher Zahlkörper. J. Fac. Sci. Univ. Tokyo, Sect. I **2**(Part 10), 477–498 (1934)

[Has34c] H. Hasse, Theorie der relativ–zyklischen algebraischen Funktionenkörper, insbesondere bei endlichem Konstantenkörper. J. Reine Angew. Math. **172**, 37–54 (1934)

[Has34d] H. Hasse, Über gewisse Ideale in einer einfachen Algebra. (Exposés mathématiques, publiés à la mémoire de Jacques Herbrand, I.). Actual. Sci. Ind. **1934**(109), 12–16 (1934)

[Has49a] H. Hasse, *Zahlentheorie* (Akademie–Verlag, Berlin, 1949), XII, 468 pp.

[Has49b] H. Hasse, Zur Frage der Zerfällungskörper des Gruppenrings einer endlichen Gruppe. Math. Nachr. **3**, 4–6 (1949)

[Has50a] H. Hasse, Invariante Kennzeichnung galoisscher Körper mit vorgegebener Galoisgruppe. Arch. Math. **2**, 281–294 (1950)

[Has50b] H. Hasse, Zum Existenzsatz von Grunwald in der Klassenkörpertheorie. J. Reine Angew. Math. **188**, 40–64 (1950)

[Has67] H. Hasse, *Vorlesungen über Klassenkörpertheorie* (Physica–Verlag, Würzburg, 1967), 275 pp.

[Has75] H. Hasse, *Mathematische Abhandlungen. Band 1, 2, 3. Herausgegeben von Heinrich Wolfgang Leopoldt und Peter Roquette* (Walter de Gruyter, Berlin, New York, 1975), Band 1: XV, 535 pp.; Band 2: XV, 525 pp.; Band 3: X, 532 pp., 1 Bild

[Has86] H. Hasse. The modern algebraic method. *Math. Intell.*, 8(2):18–25, 1986.

[Has02] H. Hasse, *Number Theory. Transl. from the 3rd German edition, edited and with a preface by Horst Günter Zimmer. Reprint of the 1980 edition* (Springer, Berlin, 2002), 638 pp.

[Hec23] E. Hecke, *Vorlesungen über die Theorie der algebraischen Zahlen* (Akad. Verlagsges., Leipzig, 1923), VIII u. 265 pp.

[Hen27] K. Hensel, Über eindeutige Zerlegung in Primelemente. J. Reine Angew. Math. **158**, 195–198 (1927)

[Her31] J. Herbrand, Sur les unités d'un corps algébrique. C. R. Acad. Sci., Paris **192**, 24–27, 188 (1931)

[Hey29] K. Hey, *Analytische Zahlentheorie in Systemen hyperkomplexer Zahlen.* Dissertation, Hamburg (1929), 49 p.

[Hil97] D. Hilbert, Die Theorie der algebraischen Zahlkörper. *Jahresber. Dtsch. Math. Ver.*, 4:I–XVIII u. 175–546, 1897. Englische Übersetzung: The Theory of Algebraic Number Fields (Springer, Heidelberg, 1998)

[Hil32] D. Hilbert, *Gesammelte Abhandlungen. Bd. 1. Zahlentheorie* (Julius Springer, Berlin, 1932), XIV, 539 pp.

[Hil33] D. Hilbert, *Gesammelte Abhandlungen. Bd. 2. Algebra. – Invariantentheorie. – Geometrie* (Julius Springer, Berlin, 1933), VIII, 453 pp., 12 Abb

[Hil35] D. Hilbert, *Gesammelte Abhandlungen. Bd. 3. Analysis. – Grundlagen der Mathematik. – Physik. – Verschiedenes. Nebst einer Lebensgeschichte* (Julius Springer, Berlin, 1935), VII, 435 pp., 12 Fig

[Iya31] S. Iyanaga, Über den allgemeinen Hauptidealsatz. Jpn. J. Math. **7**, 315–333 (1931)

[Iya34] S. Iyanaga, Zum Beweise des Hauptidealsatzes. Abh. Math. Semin. Univ. Hamb. **10**, 349–357 (1934)

[Jen86] W. Jentsch, Auszüge aus einer unveröffentlichten Korrespondenz von Emmy Noether und Hermann Weyl mit Heinrich Brandt. Historia Math. **13**, 5–12 (1986)

[Ker00] I. Kersten, Biography of Ernst Witt. Contemp. Math. **272**, 155–171 (2000)

[Kne87] M. Kneser, Max Deuring 9.12.1907 - 20.12.1984. Jahresber. Dtsch. Math.-Ver. **89**, 135–143 (1987)

[Kor36] A. Korselt, Vollständige Lösung einer neuen diophantischen Aufgabe. Math. Ann. **112**, 395–410 (1936)

[Köt30] G. Köthe, Abstrakte Theorie nichtkommutativer Ringe mit einer Anwendung auf die Darstellungstheorie kontinuierlicher Gruppen. Math. Ann. **103**, 545–572 (1930)

[Köt31] G. Köthe, Schiefkörper unendlichen Ranges über dem Zentrum. Math. Ann. **105**, 15–39 (1931)

[Köt33] G. Köthe, Erweiterung des Zentrums einfacher Algebren. Math. Ann. **107**, 761–766 (1933)

[Kru24] W. Krull, Algebraische Theorie der zerlegbaren Ringe. Math. Ann. **92**, 183–213 (1924)

[Kru28] W. Krull, Zur Theorie der allgemeinen Zahlringe. Math. Ann. **99**, 51–70 (1928)

[Kru31] W. Krull, Über die Zerlegung der Hauptideale in all-
         gemeinen Ringen. Math. Ann. **105**, 1–14 (1931)

[Kru35] W. Krull, *Idealtheorie*. Erg. d. Math. u. ihrer Grenzge-
         biete (Julius Springer, Berlin, 1935), V, 172 pp.

[KT55]   Y. Kawada, J. Tate, On the Galois cohomology of
         unramified extensions of function fields in one variable.
         Am. J. Math. **77**, 197–217 (1955)

[Lem00] F. Lemmermeyer, *Reciprocity Laws. From Euler to
         Eisenstein*. (Springer, Berlin, Heidelberg, New York,
         2000), XIX, 487 pp.

[Lev29] J. Levitzki, Über vollständig reduzible Ringe und Un-
         terringe. Nachrichten Göttingen **1929**, 240–244 (1929)

[Lev31a] J. Levitzki, A Galois theory in semi-simple rings. Bull.
         A.M.S. **37**, 44 (1931)

[Lev31b] J. Levitzki. Über vollständig reduzible Ringe und Un-
         terringe. Math. Z. **33**, 663–691 (1931)

[Lev32] J. Levitzki, On normal products of algebras.   Ann.
         Math. (2) **33**, 377–402 (1932)

[Lor70] F. Lorenz, *Quadratische Formen über Körpern*. Lec-
         ture Notes in Mathematics, vol. 130 (Springer, Berlin,
         Heidelberg, New York, 1970), I, 77 pp.

[Lor99] F. Lorenz, Nachrichten von Büchern und Menschen:
         Chiungtze C. Tsen. Sitz.ber. Math.-Nat. Kl. Akad.
         gemeinnütz. Wiss. Erfurt **9**, 1–25 (1999)

[Lor05] F. Lorenz, Käte Hey and the Main Theorem in the the-
         ory of algebras, in *European Mathematics in the Last
         Centuries*, ed. by W. Więsław, Wrocław, 2005. Stefan
         Banach International Mathematical Center, Institute
         of Mathematics Wrocław University, pp. 57–76

[Lor08] F. Lorenz, Zum Beweis der Funkionalgleichung der Heyschen Zetafunktion in der Dissertation von Wolfgang Wichmann, Emmy Noethers letztem Göttinger Doktoranden. Mitt. Math. Ges. Hamburg **27**, 167–183 (2008)

[LR] F. Lemmermeyer, P. Roquette (eds.), *Helmut Hasse and Emmy Noether. Their correspondence 1925–1935* (2006)

[LR03] F. Lorenz, P. Roquette, The theory of Grunwald-Wang in the setting of valuation theory, in *Valuation Theory and Its Applications, vol II. Proceedings of the International Conference and Workshop, University of Saskatchewan, Saskatoon, Canada, July 28–August 11, 1999*, ed. by Franz-Viktor Kuhlmann et al. Fields Inst. Commun., vol. 33 (American Mathematical Society, Providence, RI, 2003), pp. 175–212

[LR12] F. Lemmermeyer, P. Roquette (eds.), *Die mathematischen Tagebücher von Helmut Hasse 1923–1935. With an Introduction in English* (Universitäts–Verlag, Göttingen, 2012), 563 pp.

[LR16] F. Lemmermeyer, P. Roquette (eds.), *Der Briefwechsel Hasse - Scholz - Taussky* (Universitäts–Verlag, Göttingen, 2016), 574 pp.

[Mad66] M. Madan, On the Galois cohomology of tamely ramified fields of algebraic functions. Arch. Math. **17**, 400–408 (1966)

[Mag34] W. Magnus, Über den Beweis des Hauptidealsatzes. J. Reine Angew. Math. **170**, 235–240 (1934)

[ML05] S. Mac Lane, *A Mathematical Autobiography* (Karl Peters, Wellesley, MD, 2005), XV, 358 pp.

[Neu86] J. Neukirch, *Class Field Theory*. Grundlehren d. math. Wissensch. vol. 280 (Springer, Berlin, Heidelberg, New York, 1986), VIII, 140 pp.

[Noe08] E. Noether, Über die Bildung des Formensystems der ternären biquadratischen Form. J. Reine Angew. Math. **134**, 23–90 (1908)

[Noe21] E. Noether, Idealtheorie in Ringbereichen. Math. Ann. **83**, 24–66 (1921)

[Noe24] E. Noether, Abstrakter Aufbau der Idealtheorie im algebraischen Zahlkörper. Jahresber. Dtsch. Math.-Ver. **33**, 102 (1924). 2. Abteilung.

[Noe26a] E. Noether, Abstrakter Aufbau der Idealtheorie in algebraischen Zahl- und Funktionenkörpern. Math. Ann. **96**, 26–61 (1926)

[Noe26b] E. Noether, Der Endlichkeitssatz der Invarianten endlicher linearer Gruppen der Charakteristik p. Nachr. Ges. Wiss. Göttingen **1926**, 28–35 (1926)

[Noe26c] E. Noether, Gruppencharaktere und Idealtheorie. Jahresber. Dtsch. Math.-Ver. **34**(Abt.2), 144 (1926)

[Noe27] E. Noether, Der Diskriminantensatz für die Ordnungen eines algebraischen Zahl- oder Funktionenkörpers. J. Reine Angew. Math. **157**, 82–104 (1927)

[Noe29] E. Noether, Hyperkomplexe Grössen und Darstellungstheorie. Math. Z. **30**, 641–692 (1929)

[Noe32a] E. Noether, Hyperkomplexe Systeme in ihren Beziehungen zur kommutativen Algebra und Zahlentheorie. Verhandl. Intern. Math. Kongreß Zürich **1**, 189–194 (1932)

[Noe32b] E. Noether, Normalbasis bei Körpern ohne höhere Verzweigung. J. Reine Angew. Math. **167**, 147–152 (1932)

[Noe33a] E. Noether, Der Hauptgeschlechtssatz für relativ-galoissche Zahlkörper. Math. Ann. **108**, 411–419 (1933)

[Noe33b] E. Noether, Nichtkommutative Algebra. Math. Z. **37**, 514–541 (1933). Nachdruck in Ges. Abh. 40, pp. 642–669

[Noe34] E. Noether, Zerfallende verschränkte Produkte und ihre Maximalordnungen. (Exposés mathématiques, publiés à la mémoire de Jacques Herbrand, IV.). Actual. Sci. Ind. **1934**(148), 15 p. (1934)

[Noe50] E. Noether, Idealdifferentiation und Differente. J. Reine Angew. Math. **188**, 1–21 (1950)

[Noe83a] E. Noether, Algebra der hyperkomplexen Größen. Vorlesung Wintersemester 1929/30, ausgearbeitet von M. Deuring, in *Emmy Noether, Collected Papers*, ed. by N. Jacobson (Springer, New York, 1983), pp. 711–763, VIII, 777 pp.

[Noe83b] E. Noether, Gesammelte Abhandlungen. Collected papers. Hrsg. von N. Jacobson (Springer, Berlin, Heidelberg, New York, Tokyo), VIII, 777 S. DM 144.00; $ 64.00 (1983)

[Ost19] A. Ostrowski, Zur arithmetischen Theorie der algebraischen Grössen. Nachr. Göttinger Gesellsch. d. Wiss. **1919**, 279–298 (1919)

[Pet35] K. Petri, Über die Diskriminante ternärer Formen. Sitzungsber. Bayer. Akad. Wiss., Math.—Naturwiss. Abt. **1935**(3), 471–484 (1935)

[Pfi65] A. Pfister, Zur Darstellung von -1 als Summe von Quadraten in einem Körper. Sitzungsber. Bayer. Akad. Wiss., Math.—Naturwiss. Abt. **1965**(40), 159–165 (1965)

[Pie31a] S. Pietrkowski, Theorie der unendlichen Abelschen Gruppen. Math Ann. **104**, 535–569 (1931)

[Pie31b]   S. Pietrkowski, Untergruppen und Quotientengruppen unendlicher Abelscher Gruppen. Math. Ann. **105**, 666–671 (1931)

[Prü25]   H. Prüfer, Neue Begründung der algebraischen Zahlentheorie. Math. Ann. **94**, 198–243 (1925)

[Rau26]   H. Rauter, *Über die Darstellbarkeit durch quadratische Formen im Körper der rationalen Funktionen einer Unbestimmten über dem Restklassenkörper mod p.* PhD thesis, Universität Halle, 1926

[Rau28a]   H. Rauter, Bemerkung zu der Arbeit: "Studien zur Theorie des Galoisschen Körpers...". J. Reine Angew. Math. **159**, 228 (1928)

[Rau28b]   H. Rauter, Studien zur Theorie des Galoisschen Körpers über dem Körper der rationalen Funktionen einer Unbestimmten t mit Koeffizienten aus einem beliebigen endlichen Körper von p^{m_0} Elementen. J. Reine Angew. Math. **159**, 117–132 (1928)

[Rei33]   H. Reichardt, Arithmetische Theorie der kubischen Körper als Radikalkörper. Monatsh. Math. Phys. **40**, 323–350 (1933)

[Roq01]   P. Roquette, Class field theory in characteristic p, its origin and development, in *Class Field Theory – Its Centenary and Prospect. Proceedings of the 7th MSJ International Research Institute of the Mathematical Society of Japan, Tokyo, Japan, June 3–12, 1998*, ed. by K. Miyake, vol. 30. Adv. Stud. Pure Math. (Mathematical Society of Japan, Tokyo, 2001), pp. 549–631

[Roq05a]   P. Roquette, *The Brauer-Hasse-Noether Theorem in historical perspective.* Schriftenreihe der Heidelberger Akademie der Wissenschaften, vol. 15 (Springer, Berlin, Heidelberg, New York, 2005), I, 77 pp.

[Roq05b] P. Roquette, *The Brauer-Hasse-Noether Theorem in Historical Perspective* (Springer, Berlin, 2005)

[Roq08] P. Roquette, Über Wolfgang Wichmann. Preprint (2008), 4 pp.

[Roq10] P. Roquette, *Contributions to the History of Number Theory in the 20th Century* (European Mathematical Society (EMS), Zürich, 2010), 278 pp.

[Roq14] P. Roquette, Jacques Herbrand und sein Lemma. Mitt. Math. Ges. Hamburg. **34**, 163–194 (2014)

[Roq18] P. Roquette, *The Riemann Hypothesis in Characteristic p in Historical Perspective*. Lecture Notes in Mathematics, vol. 2222 (Springer, New York, 2018)

[Sch06] I. Schur, Arithmetische Untersuchungen über endliche Gruppen linearer Substitutionen. Sitz. Ber. Preuss. Akad. Wiss. Berlin **1906**, 164–184 (1906)

[Sch28] F.K. Schmidt, Über die Primzahlzerlegung der Hauptideale eines Integritätsbereichs. Sitzungsberichte München **1928**, 285–287 (1928)

[Sch30] F.K. Schmidt, Zur Klassenkörpertheorie im Kleinen. J. Reine Angew. Math. **162**, 155–168 (1930)

[Sch31a] F.K. Schmidt, Analytische Zahlentheorie in Körpern der Charakteristik p. Math. Z. **33**, 1–32 (1931)

[Sch31b] F.K. Schmidt, Die Theorie der Klassenkörper über einem Körper algebraischer Funktionen in einer Unbestimmten und mit endlichem Koeffizientenbereich. Sitzungsber. Physik.-Med. Soz. Erlangen **62**, 267–284 (1931)

[Sch35a] O.F.G Schilling, Über gewisse Beziehungen zwischen der Arithmetik hyperkompler Zahlsysteme und algebraischer Zahlkörper. Math. Annalen **111**, 372–398 (1935)

[Sch35b] H.L. Schmid, Über das Reziprozitätsgesetz in relativ–zyklischen algebraischen Funktionenkörpern mit endlichem Konstantenkörper. Math. Z. **40**, 91–109 (1935)

[Sch87] N. Schappacher, Das mathematische Institut der Universität Göttingen 1929–1950, in *Die Universität Göttingen unter dem Nationalsozialismus*, ed. by H. Becker Andere (K. G. Saur, Munich, 1987), pp. 345–373

[Sch27] F.K. Schmidt, Zur Zahlentheorie in Körpern von der Charakteristik p. Sitzungsberichte Erlangen **58/59**, 159–172 (1926/27)

[Seg03] S.L. Segal, Mathematicians Under the Nazis (Princeton University Press, Princeton, NJ, 2003), xxii, 530 pp.

[Sho32] K. Shoda, Über die Galoissche Theorie der halbeinfachen hyperkomplexen Systeme. Math. Ann. **107**, 252–258 (1932)

[Spe16] A. Speiser, Gruppendeterminante und Körperdiskriminante. Math. Ann. **77**, 546–562 (1916)

[Spe19] A. Speiser, Zahlentheoretische Sätze aus der Gruppentheorie. Math. Z. **5**, 1–6 (1919)

[Spe27] A. Speiser, *Die Theorie der Gruppen von endlicher Ordnung. Mit Anwendungen auf algebraische Zahlen und Gleichungen sowie auf die Kristallographie.* Die Grundlehren der mathematischen Wissenschaften mit besonderer Berücksichtigung ihrer Anwendungsgebiete Bd. 5, 2nd edn. (J. Springer, Berlin, 1927), IX + 251 pp. mit 38 Abb

[Sta36] R. Stauffer, The construction of a normal basis in a separable normal extension field. Am. J. Math. **58**, 585–597 (1936)

[Ste10] E. Steinitz, Algebraische Theorie der Körper. J. Reine Angew. Math. **137**, 167–309 (1910)

[Ste30] E. Steinitz, *Algebraische Theorie der Körper. Neu herausgegeben, mit Erläuterungen und einem Anhang: Abriß der Galoisschen Theorie versehenen von R. Baer und H. Hasse* (de Gruyter-Verlag, Berlin, 1930), 177 pp.

[Tak20] T. Takagi, Über eine Theorie des relativ abelschen Zahlkörpers. J. College of Science, Imp. Univ. of Tokyo. **41**, 1–133 (1920). In Collected Papers, 13., pp. 73–167

[Tob03] R. Tobies, Briefe Emmy Noethers an P.S. Alexandroff. NTM N.S. **11**(2), 100–115 (2003)

[Toe49] O. Toeplitz, *Die Entwicklung der Infinitesimal-rechnung. Eine Einleitung in die Infinitesimalrech-nung nach der genetischen Methode. Erster Band. Aus dem Nachlass herausgegeben von G. Köthe.* Die Grundlehren der mathematischen Wissenschaften in Einzeldarstellungen. LVI (Springer, Berlin, Göttingen, Heidelberg, 1949), IX, 181 pp., 148 Abb

[Tol] C. Tollmien, *Die Lebens- und Familiengeschichte der Mathematikerin Emmy Noether in Einzelaspekten* (2021)

[Tse33] C. Tsen, Divisionsalgebren über Funktionenkörpern. Nachr. Ges. Wiss. Göttingen, Math.—Phys. Kl. I **1933**(44), 335–339 (1933)

[Tse34] C. Tsen, *Algebren über Funktionenkörpern.* Disserta-tion, Göttingen, 1934. 19 p.

[TT81] O. Taussky-Todd, My personal recollections of Emmy Noether, in *Emmy Noether. A Tribute to Her Life and Work*, ed. by J.W. Brewer, M.K. Smith (M. Dekker, New York, 1981), pp. 79–92

[Ulm33] H. Ulm, Zur Theorie der abzählbar–unendlichen Abelschen Gruppen. Math. Ann. **107**, 774–803 (1933)

[Ulm35] H. Ulm, Zur Theorie der nicht–abzählbaren primären abelschen Gruppen. Math. Z. **40**, 205–207 (1935)

[vdW30] B.L. van der Waerden, *Moderne Algebra. Unter Benutzung von Vorlesungen von E. Artin und E. Noether. Bd. I.* Die Grundlehren der mathematischen Wissenschaften in Einzeldarstellungen mit besonderer Berücksichtigung der Anwendungsgebiete Bd. 23 (Springer, Berlin, 1930), VIII + 243 pp.

[vdW31] B.L. van der Waerden, *Moderne Algebra. Unter Benutzung von Vorlesungen von E. Artin und E. Noether. Bd. II.* Die Grundlehren der mathematischen Wissenschaften in Einzeldarstellungen mit besonderer Berücksichtigung der Anwendungsgebiete Bd. 24 (Springer, Berlin, 1931), VII + 216 pp.

[vdW34] B.L. van der Waerden, Elementarer Beweis eines zahlentheoretischen Existenztheorems. J. Reine Angew. Math. **171**, 1–3 (1934)

[vdW35] B.L. van der Waerden, Nachruf auf Emmy Noether. Math. Ann. **111**, 469–476 (1935)

[vdW75] B.L. van der Waerden, On the sources of my book Moderne Algebra. Historia Math. **2**, 11–40 (1975)

[Ven70] B.A. Venkov, *Elementary Number Theory.* Wolters–Noordhoff Series of Monographs and Textbooks on Pure and Applied Mathematics (Wolters–Noordhoff Publishing, Groningen, The Netherlands, 1970), IX, 249 pp.

[vN26] J.v. Neumann, Zur Prüferschen Theorie der idealen Zahlen. Acta Szeged **2**, 193–227 (1926)

[Vor35] W. Vorbeck, *Nichtgaloissche Zerfällungskörper ein-*
         facher hyperkomplexer Systeme. PhD thesis, Universiät
         Göttingen, 1935

[Wan48] S. Wang, A counter example to Grunwald's theorem.
         Ann. Math. (2) **49**, 1008–1009 (1948)

[Wan50] S. Wang, On Grunwald's theorem. Ann. Math. **51**,
         471–484 (1950)

[Web12] H. Weber, *Lehrbuch der Algebra. Kleine Ausgabe in*
         einem Bande (Friedr. Vieweg u. Sohn, Braunschweig,
         1912), X, 528 pp.

[Wed14] I.H.M. Wedderburn, A type of primitive algebra.
         Trans. Am. Math. Soc. **15**, 162–166 (1914)

[Wei93] A. Weil, *Lehr- und Wanderjahre eines Mathematikers.*
         Aus dem Französischen übersetzt von Theresia Übelhör
         (Birkhäuser, Basel, 1993), 212 pp.

[Wen31] B.A. Wenkov, Über die Klassenzahl positiver binärer
         quadratischer Formen. Math. Z. **33**, 350–374 (1931)

[Wey35] H. Weyl, Emmy Noether. Scripta math. **3**, 201–220
         (1935). Reprinted in the Noether biography of Auguste
         Dick 1970

[Wit31] E. Witt,   Über die Kommutativität endlicher
         Schiefkörper. Abh. Math. Semin. Univ. Hamb. **8**, 413
         (1931)

[Wit34a] E. Witt, Riemann–Rochscher Satz und ζ–Funktion im
         Hyperkomplexen. Math. Ann. **110**, 12–28 (1934)

[Wit34b] E. Witt, Über ein Gegenbeispiel zum Normensatz.
         Math. Z. **39**, 462–467 (1934)

[Wit34c] E. Witt, Zerlegung reeller algebraischer Funktionen in
         Quadrate. Schiefkörper über rellem Funktionenkörper.
         J. Reine Angew. Math. **171**, 1–11 (1934)

[Wit35a] E. Witt, Der Existenzsatz für abelsche Funktionenkörper. J. Reine Angew. Math. **173**, 43–51 (1935)

[Wit35b] E. Witt, Zwei Regeln über verschränkte Produkte. J. Reine Angew. Math. **173**, 191–192 (1935)

[Wit36a] E. Witt, Bemerkungen zum Beweis des Hauptidealsatzes von S. Iyanaga. Abh. Math. Semin. Univ. Hamb. **11**, 221 (1936)

[Wit36b] E. Witt, Zyklische Körper und Algebren der Charakteristik p vom Grad p^n. Struktur diskret bewerteter perfekter Körper mit vollkommenem Restklassenkörper der Charakteristik p. J. Reine Angew. Math. **176**, 126–140 (1936)

[Wit37] E. Witt, Theorie der quadratischen Formen in beliebigen Körpern. J. Reine Angew. Math. **176**, 31–44 (1937)

[Wit54] E. Witt, Verlagerung von Gruppen und Hauptidealsatz, in *Proceedings of the ICM 1954*, vol. 2 (International Mathematical Union, North-Holland, Amsterdam, 1954), pp. 71–73

[Wit98] E. Witt, *Collected papers – Gesammelte Abhandlungen*, ed. by Ina Kersten. With an essay by Günter Harder on Witt vectors (Springer, Berlin, 1998), xvi, 420 pp.

[Zas37] H. Zassenhaus, *Lehrbuch der Gruppentheorie. Bd. 1.*. Hamburg. Math. Einzelschriften, vol. 21 (B. G. Teubner, Leipzig, Berlin, 1937), VI, 152 pp.

[Zor33] M. Zorn, Note zur analytischen hyperkomplexen Zahlentheorie. Abh. Math. Semin. Univ. Hamb. **9**, 197–201 (1933)

Index

© Springer Nature Switzerland AG 2022
P. Roquette, F. Lemmermeyer, *The Hasse - Noether Correspondence 1925 - 1935*, History of Mathematics Subseries 2317,
https://doi.org/10.1007/978-3-031-12880-6

LECTURE NOTES IN MATHEMATICS  Springer

Editors in Chief: J.-M. Morel, B. Teissier;

Editorial Policy

1. Lecture Notes aim to report new developments in all areas of mathematics and their applications – quickly, informally and at a high level. Mathematical texts analysing new developments in modelling and numerical simulation are welcome.

   Manuscripts should be reasonably self-contained and rounded off. Thus they may, and often will, present not only results of the author but also related work by other people. They may be based on specialised lecture courses. Furthermore, the manuscripts should provide sufficient motivation, examples and applications. This clearly distinguishes Lecture Notes from journal articles or technical reports which normally are very concise. Articles intended for a journal but too long to be accepted by most journals, usually do not have this "lecture notes" character. For similar reasons it is unusual for doctoral theses to be accepted for the Lecture Notes series, though habilitation theses may be appropriate.

2. Besides monographs, multi-author manuscripts resulting from SUMMER SCHOOLS or similar INTENSIVE COURSES are welcome, provided their objective was held to present an active mathematical topic to an audience at the beginning or intermediate graduate level (a list of participants should be provided).

   The resulting manuscript should not be just a collection of course notes, but should require advance planning and coordination among the main lecturers. The subject matter should dictate the structure of the book. This structure should be motivated and explained in a scientific introduction, and the notation, references, index and formulation of results should be, if possible, unified by the editors. Each contribution should have an abstract and an introduction referring to the other contributions. In other words, more preparatory work must go into a multi-authored volume than simply assembling a disparate collection of papers, communicated at the event.

3. Manuscripts should be submitted either online at www.editorialmanager.com/lnm to Springer's mathematics editorial in Heidelberg, or electronically to one of the series editors. Authors should be aware that incomplete or insufficiently close-to-final manuscripts almost always result in longer refereeing times and nevertheless unclear referees' recommendations, making further refereeing of a final draft necessary. The strict minimum amount of material that will be considered should include a detailed outline describing the planned contents of each chapter, a bibliography and several sample chapters. Parallel submission of a manuscript to another publisher while under consideration for LNM is not acceptable and can lead to rejection.

4. In general, **monographs** will be sent out to at least 2 external referees for evaluation.

A final decision to publish can be made only on the basis of the complete manuscript, however a refereeing process leading to a preliminary decision can be based on a pre-final or incomplete manuscript.

Volume Editors of **multi-author works** are expected to arrange for the refereeing, to the usual scientific standards, of the individual contributions. If the resulting reports can be forwarded to the LNM Editorial Board, this is very helpful. If no reports are forwarded or if other questions remain unclear in respect of homogeneity etc, the series editors may wish to consult external referees for an overall evaluation of the volume.

5. Manuscripts should in general be submitted in English. Final manuscripts should contain at least 100 pages of mathematical text and should always include

- a table of contents;
- an informative introduction, with adequate motivation and perhaps some historical remarks: it should be accessible to a reader not intimately familiar with the topic treated;
- a subject index: as a rule this is genuinely helpful for the reader.
- For evaluation purposes, manuscripts should be submitted as pdf files.

6. Careful preparation of the manuscripts will help keep production time short besides ensuring satisfactory appearance of the finished book in print and online. After acceptance of the manuscript authors will be asked to prepare the final LaTeX source files (see LaTeX templates online: https://www.springer.com/gb/authors-editors/book-authors-editors/manuscriptpreparation/5636) plus the corresponding pdf- or zipped ps-file. The LaTeX source files are essential for producing the full-text online version of the book, see http://link.springer.com/bookseries/304 for the existing online volumes of LNM). The technical production of a Lecture Notes volume takes approximately 12 weeks. Additional instructions, if necessary, are available on request from lnm@springer.com.

7. Authors receive a total of 30 free copies of their volume and free access to their book on SpringerLink, but no royalties. They are entitled to a discount of 33.3 % on the price of Springer books purchased for their personal use, if ordering directly from Springer.

8. Commitment to publish is made by a *Publishing Agreement*; contributing authors of multiauthor books are requested to sign a *Consent to Publish form*. Springer-Verlag registers the copyright for each volume. Authors are free to reuse material contained in their LNM volumes in later publications: a brief written (or e-mail) request for formal permission is sufficient.

Addresses:

Professor Jean-Michel Morel, CMLA, École Normale Supérieure de Cachan, France
E-mail: moreljeanmichel@gmail.com

Professor Bernard Teissier, Equipe Géométrie et Dynamique,
Institut de Mathématiques de Jussieu – Paris Rive Gauche, Paris, France
E-mail: bernard.teissier@imj-prg.fr

Springer: Ute McCrory, Mathematics, Heidelberg, Germany,
E-mail: lnm@springer.com